D0430532

LUCY'S LEGACY

LUCY'S LEGACY

The Quest for Human Origins

Donald C. Johanson

Kate Wong

THREE RIVERS PRESS
NEW YORK

Published in the United States by Three Rivers Press, an imprint of the
Crown Publishing Group, a division of Random House, Inc., New York.
www.crownpublishing.com

Three Rivers Press and the Tugboat design are registered trademarks of Random House, Inc.
Originally published in hardcover in slightly different form in the United States by
Harmony Books, an imprint of the Crown Publishing Group, a division of
Random House, Inc., New York, in 2009.

Library of Congress Cataloging-in-Publication Data

Johanson, Donald C.
Lucy's legacy / Donald Johanson and Kate Wong.—1st ed.
p. cm.
1. Lucy (Prehistoric hominid).   2. *Australopithecus afarensis.*
3. Human beings—Origin.   I. Wong, Kate.   II. Title.
GN283.25.J64 2009
569.9—dc22
2008039907

ISBN 978-0-307-39640-2

Printed in the United States of America

Design by Leonard W. Henderson

The Hominid Family Tree by Michael Hagelberg

Key Hominid Sites map and Hadar Formation graph by Mapping Specialists

10   9   8   7   6   5   4   3   2   1

First Paperback Edition

To my late mentor Paul Leser,
for precious understanding and
enduring inspiration.
—D.C.J.

To my parents, Ann and C.C.,
for their unflagging support.
—K.W.

Contents

LUCY'S LEGACY

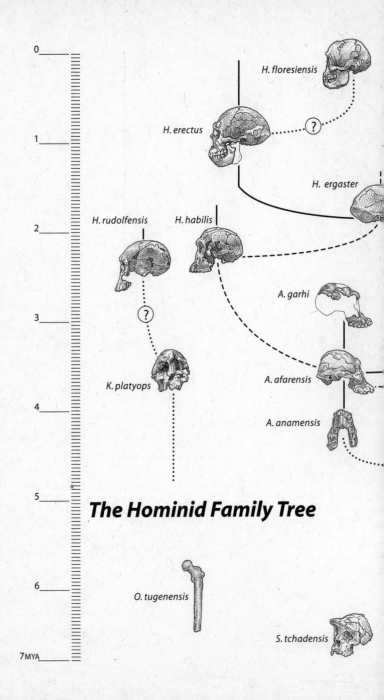

0

1

2

3

4

5

6

7 MYA

H. floresiensis

H. erectus

H. ergaster

H. rudolfensis

H. habilis

?

A. garhi

A. afarensis

A. anamensis

?

K. platyops

The Hominid Family Tree

O. tugenensis

S. tchadensis

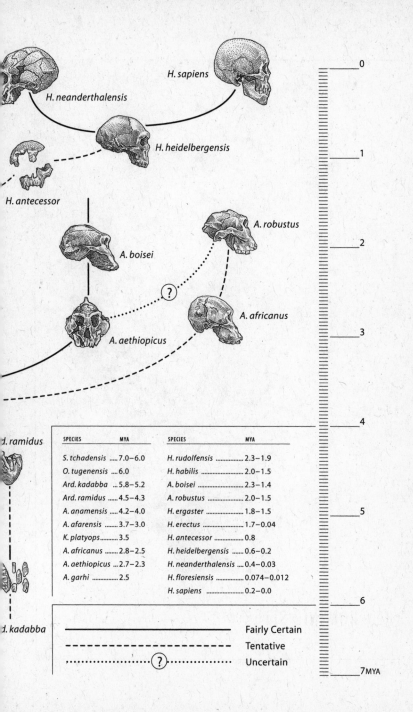

H. sapiens

H. neanderthalensis

H. heidelbergensis

H. antecessor

A. robustus

A. boisei

?

A. africanus

A. aethiopicus

0

1

2

3

4

H. ramidus

SPECIES	MYA	SPECIES	MYA
S. tchadensis	7.0–6.0	H. rudolfensis	2.3–1.9
O. tugenensis	6.0	H. habilis	2.0–1.5
Ard. kadabba	5.8–5.2	A. boisei	2.3–1.4
Ard. ramidus	4.5–4.3	A. robustus	2.0–1.5
A. anamensis	4.2–4.0	H. ergaster	1.8–1.5
A. afarensis	3.7–3.0	H. erectus	1.7–0.04
K. platyops	3.5	H. antecessor	0.8
A. africanus	2.8–2.5	H. heidelbergensis	0.6–0.2
A. aethiopicus	2.7–2.3	H. neanderthalensis	0.4–0.03
A. garhi	2.5	H. floresiensis	0.074–0.012
		H. sapiens	0.2–0.0

5

6

H. kadabba

————————————————————— Fairly Certain

– – – – – – – – – – – – – – Tentative

···············(?)··············· Uncertain

7 MYA

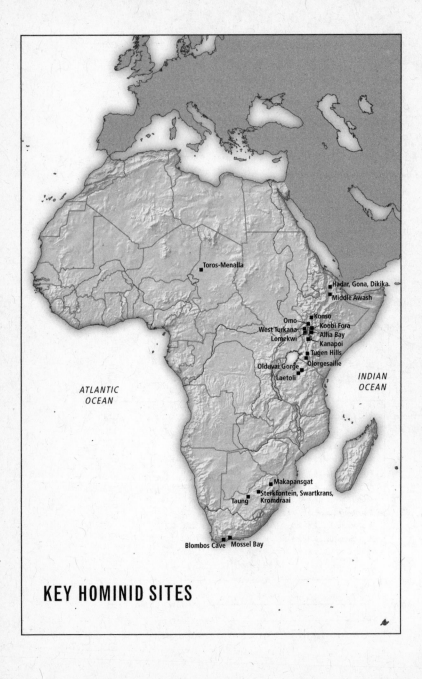

Toros-Menalla

Hadar, Gona, Dikika.
Middle Awash

Konso
Omo
West Turkana   Koobi Fora
Lomekwi   Allia Bay
Kanapoi
Tugen Hills
Olduvai Gorge   Olorgesailie
Laetoli

ATLANTIC
OCEAN

INDIAN
OCEAN

Makapansgat
Sterkfontein, Swartkrans,
Kromdraai
Taung

Blombos Cave   Mossel Bay

KEY HOMINID SITES

PART 1

Lucy

The Woman Who Shook Up
Man's Family Tree

Never in my wildest fantasies did I imagine that I would discover a fossil as earthshaking as Lucy. When I was a teenager, I dreamed of traveling to Africa and finding a "missing link." Lucy is that and more: a 3.2-million-year-old skeleton who has become the spokeswoman for human evolution. She is perhaps the best known and most studied fossil hominid of the twentieth century, the benchmark by which other discoveries of human ancestors are judged.

Whenever I tell the story, I am instantly transported back to the thrilling moment when I first saw her thirty-four years ago on the sandy slopes of Hadar in Ethiopia's Afar region. I can feel the searing, noonday sun beating down on my shoulders, the beads of sweat on my forehead, the dryness of my mouth—and then the shock of seeing a small fragment of bone lying inconspicuously on the ground. Most dedicated fossil hunters spend the majority of their lives in the field without finding anything remarkable, and there I was, a thirty-one-year-old newly minted Ph.D., staring at my childhood dream at my feet.

Sunday, November 24, 1974, began, as it usually does for me in the field, at dawn. I had slept well in my tent, with the glittering stars visible through the small screen that kept out the mosquitoes, and as sunrise announced a brilliant new day, I got up and went to the dining tent for a cup of thick, black Ethiopian coffee. Listening to the morning sounds of camp life, I planned with some disinclination the day's activities: catching up on correspondence, fossil cataloging, and a million other tasks that had been set aside to accommodate a visit from anthropologists Richard and Mary Leakey. I looked up as Tom Gray, my grad student, appeared.

"I'm plotting the fossil localities on the Hadar map," he said. "Can you show me Afar Locality 162, where the pig skull was found last year?"

"I have a ton of paperwork and am not sure I want to leave camp today."

"Can you do the paperwork later?"

"Even if I start it now I'll be doing it later," I grumbled. But something inside—a gut sense that I had learned to heed—said I should put the paperwork aside and head to the outcrops with Tom.

A couple of geologists joined us in one of our old, dilapidated Land Rovers, and in a cloud of dust we headed out to the field. I sat in the passenger seat enjoying the passing landscape peppered with animal fossils. Flocks of quacking guinea fowl ran for cover, and a giant warthog, annoyed by our intrusion, hurried off, its tail straight up in the air. Unlike many mammals that had been hunted to extinction in the area, the Hadar warthogs were left alone by the Afar locals, whose Islamic faith forbade eating pork. Tom put the Land Rover through its paces, and as we picked up speed in the sandy washes, my mind switched gears into fossil-finding mode. After we dropped off the geologists, who needed to inspect a troublesome geological fault that had disturbed the sedimentary layers near Locality 162, Tom and I threaded our way along smaller and smaller gullies.

"Somewhere around here," I said. "Pull over." Then I laughed as it occurred to me that in the remote desert you don't have to pull over, you just stop driving. We got out and spent a few minutes locating the cairn that had been left to mark the pig skull's locality, a little plateau of clay and silt sediments bordered by harder layers of sandstone. A year earlier, a geologist had been out on a mapping mission and the plateau was obvious on the aerial photographs we had toted along; otherwise we might have overlooked it. After carefully piercing a pinhole into the aerial photo to mark the spot and labeling it "162" on the reverse side, we lingered. I was reluctant to return to camp and my paperwork. Even though the area was known to be fossil poor, we decided to look around while we were there. But after two hours of hunting all we had to show were some unremarkable fossil antelope and horse teeth, a bit of a pig skull, and a fragment of monkey jaw.

"I've had it. When do we head back?" Tom said.

"Right now." With my gaze still glued to the ground, I cut across the midportion of the plateau toward the Land Rover. Then a glint caught my eye, and when I turned my head I saw a two-inch-long, light brownish gray fossil fragment shaped like a wrench, which my knowledge of osteology told me instantly was part of an elbow. I knelt and picked it up for closer inspection. As I examined it, an image clicked into my brain and a subconscious template announced hominid. (The term *hominid* is used throughout this book to refer to the group of creatures in the human lineage since they diverged from a common ancestor to the African apes. Some other scholars employ the word *hominin* in its place.) The only other thing it could have been was monkey, but it lacked the telltale flare on the back that characterizes monkey elbows. Without a doubt, this was the elbow end of a hominid ulna, the larger of the two bones in the forearm. Raising my eyes, I scanned the immediate surroundings and spotted other bone fragments of similar color—a piece of thighbone, rib fragments, segments of the backbone, and, most important, a shard of skull vault.

"Tom, look!" I showed him the ulna, then pointed at the fragments. Like me, he dropped to a crouch. With his jaw hanging open, he picked up a chunk of mandible that he wordlessly held out for me to see. "Hominid!" I gushed. "All hominid!" Our excitement mounted as we examined every splinter of bone. "I don't believe this! Do you believe this?" we shouted over and over. Drenched in sweat, we hugged each other and whooped like madmen.

"I'm going to bring the ulna to camp," I said. "We'll come back for the others." I wanted to mark the exact location of each bone fragment scattered on the landscape, but there were too many pieces and time was short.

"Good idea. Don't lose it," Tom joked, as I carefully wrapped the ulna in my bandanna. I decided to take a fragment of lower jaw, too, for good measure. I marked the exact spots where the bones had lain, scribbled a few words in my field notebook, and then got back into the Land Rover.

The two geologists relaxing in the shade of a small acacia tree looked relieved when we drove up to rescue them from the stultifying heat. As they stood and greeted us, they could tell from our giddy grins that we'd found something.

"Feast your eyes!" I said, and opened the bandanna. I held the ulna next to my elbow. Being geologists, they didn't know a lot about bones, but they understood the importance of the find. Back into the bandanna the bones went, and then into my khaki hat for the trip to camp in the safety of my lap. Thirty minutes later Tom announced our arrival by honking the horn, and as we pulled to a stop our inquisitive teammates surrounded the car.

I jumped out of the Land Rover and everyone followed me to the work area, where a large tent fly protected our plywood worktables. Still in a state of semidisbelief, I sat and unpacked the precious remains. Reassured that they were in fact real, I sighed with relief. Everyone leaned over to see the tiny fragments of arm and jaw. The questions came fast and furious. Is there more? Where'd you find it? How did you find it? And then there was a stunned silence as the import of what we'd found sunk in. It hit me that if I had walked just a few more paces and looked to my left rather than my right, the bones would still be there on the slope. And in the ever-changing landscape of the Afar, a single desert thunderstorm could have washed them off the plateau, over a cliff and into oblivion, forever.

Suddenly someone slapped me on the back and exhilaration replaced awe. We all started talking at once, and we had to keep raising our voices to be heard so that eventually no one could hear what anyone else was saying. A hurried lunch followed and then everyone wanted to see the spot where I had found the ulna. At the locality my colleagues stood back as I carefully pointed to the bone fragments on the slope. Immediately my team understood that what they were looking at was a partial hominid skeleton. It was a special moment for all of us, though I don't think any of us truly realized how special at the time.

We celebrated the discovery with a delicious dinner of roasted goat and panfried potatoes washed down with a case of Bati beer my students had somehow managed to smuggle into camp. Conversation became less animated and more technical, focusing on morphology and size. I felt from the beginning that the fossils belonged to a single individual because there was no duplication of parts in the remains we collected; the pieces all had the same proportions and exhibited the same fossilization color. I further argued that the skeleton was a female specimen of *Australopithecus*—a primitive human forebear—because of the small size of the bones relative to those of other australopithecines. All australopithecines were sexually dimorphic, which is to say males and females exhibited physical differences beyond those pertaining to the sex organs. So if the lightly built ulna we discovered were from a male, then a female would have to be unbelievably tiny.

While we were all talking, *Sgt. Pepper's Lonely Hearts Club Band* was playing on a small Sony tape deck. When "Lucy in the Sky with Diamonds" came on, my girlfriend Pamela Alderman, who had come to spend some time in the field with me, said, "Why don't you call her Lucy?" I smiled politely at the suggestion, but I didn't like it because I thought it was frivolous to refer to such an important find simply as Lucy. Nicknaming hominid fossils was not unheard of, however. Mary and Louis Leakey, giants in the field of paleoanthropology, dubbed a flattened hominid skull found in Tanzania's Olduvai Gorge "Twiggy," and a specimen their son Jonathan found received the moniker "Jonny's Child." But most of the scientists I knew wouldn't give their fossils a cute name based on a song by the Beatles. The next morning, however, everyone wanted to know if we were going to the Lucy site. Someone asked how tall Lucy was. Another inquired how old I thought Lucy was when she died. As I sat there eating my breakfast of peanut butter and jelly on toast, I conceded that the name Lucy had a better ring to it than A.L. 288, the locality number that had been assigned to the site.

At my request, the government representative from the Antiquities Administration who had escorted our expedition sent word to the director general of the Ministry of Culture, Bekele Negussie. He arrived a few days later with some of his colleagues. While I answered their questions, I resisted referring to our australopithecine as Lucy because I was uncomfortable about an Ethiopian fossil bearing an English name. When the team returned that afternoon from the site bursting with news of more Lucy fragments, additional information about Lucy, endless speculations about Lucy, my discomfort grew. After dinner Bekele and I sat outside the dining tent looking up at a brilliant starlit sky. I talked about the implications of the discovery, how it might impact prevailing theories about hominid evolution. And we discussed arrangements for a press announcement in Addis Ababa in December.

He listened in silence, then regarded me very seriously and said, "You know, she is an Ethiopian. She needs an Ethiopian name."

"Yes!" I agreed, relieved. "What do you suggest?"

"Dinkinesh is the perfect name for her."

I mentally inventoried my Amharic vocabulary, which was just enough to shop for basics, greet people, ask directions, and, most important, order a cup of the best coffee in the world. The word *Dinkinesh* wasn't there. "What does it mean?"

With a broad smile, as if he were naming his own child, he answered, "Dinkinesh means 'you are marvelous.'"

He was right, it was the perfect name. Of course, today most of the world, including nearly every Ethiopian I have spoken to, calls her Lucy. And Lucy is the name that has appeared in crossword puzzles, on *Jeopardy!*, in cartoons, and on African Red Bush Tazo tea bags. In Ethiopia she has lent her name to numerous coffee shops, a rock band, a typing school, a fruit juice bar, and a political magazine. There is even an annual Lucy Cup soccer competition in Addis Ababa. Once, while driving through the town of Kombolcha on the way back to Addis after a field season, years after the discovery, I spotted a small sign that said LUSSY BAR. I brought the car to a screeching halt and my colleagues and I went in to have a beer. When we asked the proprietress how the place got its name, she explained in a solemn voice that many years ago a young American found a skeleton named Lucy in the Afar region, and that she took great pride in naming her bar after the fossil that proved Ethiopia's status as the original homeland to all people. With a grin, I told her I was the American who had found Lucy. She shrieked in delight and insisted that we have our picture taken together to mount on the wall. I sent the photo to her, and for all I know, it hangs there still. But sometimes I still think of Lucy as Dinkinesh, because she truly is marvelous.

At the end of the 1974 field season, Lucy, painstakingly wrapped and packed in a cardboard box, made the day-and-a-half journey from Hadar to the National Museum of Ethiopia in Addis Ababa. Lucy was expected, and when my colleagues and I pulled up to the museum in my Land Rover, the director, Mamo Tessema, and his curatorial staff greeted us warmly. The cardboard box, a temporary home for Lucy, was whisked off and locked in a carefully guarded room.

Over the next few days I worked with Woldesenbet, the no-nonsense collections manager, to officially tally every fragment and formally catalog Lucy as part of the National Museum of Ethiopia's collections. The entire Ministry of Culture was abuzz with the news that she was now in Addis Ababa, and Bekele and the minister made preparations for her official coming-out party, a public announcement at the ministry. A throng of scientists, antiquities people, ministry officials, university faculty, and journalists jammed the room that had been specially prepared by the ministry. Resplendent on a black cloth, Lucy was an instant hit. Everyone in the room jostled to catch a glimpse of her bones.

The press conference was intense, and Bekele had to finally call a stop to the questions. Three million two hundred thousand years after her death her image would grace the front page of newspapers all over the world. Not exactly an overnight success, I thought. In a way, we had both arrived, Lucy and I. Her path to that press conference was quite different from mine, to be sure. But the reason we were there together was because we had met in the right place at the right time.

My own path could be characterized as a combination of hard work, perseverance, and more than a few lucky breaks. My first opportunity to travel to Africa had come four years earlier, in 1970, when one of my professors at the University of Chicago, F. Clark Howell, invited me to work as an assistant on his expedition to the Omo area in extreme southwestern Ethiopia, where he was looking for hominids. This was the chance I had been dreaming of ever since I had read as a kid the *National Geographic* articles describing the Leakeys' finds at Olduvai Gorge. I was utterly mesmerized by their fascinating fieldwork and the photos of Africa. But when I stepped out of the light plane at the Omo airstrip I was astounded by the brutal heat as I staggered to the waiting Land Rover. Soon I suffered an attack of malaria that could have quickly terminated my career as a field scientist, but eventually I became acclimated to the rigors of working in Africa. During my first expedition, the weeks passed swiftly as I grew more and more comfortable in the field and learned many of the skills necessary to conduct fieldwork. After three months I was hooked, and I knew that Africa would forever be a part of my life.

In the summer of 1971 I returned to the Omo and was given greater responsibility in running the camp alongside Gerry Eck, our camp manager extraordinaire. More than anything in the world I wanted an expedition of my own someday, and I knew that all the work I was doing on the Omo Research Expedition was the best preparation I could ever have.

In the fall of 1971 I spent several weeks in Europe gathering data for my doctoral dissertation. I was studying variation in chimpanzee teeth, which could inform assessments of similar variation in early hominids. My time working in Paris proved to be the most memorable and rewarding of my young career, for many reasons. From my experience in the Omo I knew several members of the French contingent, who looked after me in Paris. One evening I attended a dinner party where I met a young Tunisian-born French geologist named Maurice Taieb, a gregarious, handsome, and heavily tanned man with a highly animated personality. As he sipped a glass of red wine and took deep drags off one Gauloise after another, he regaled me with stories of his fieldwork in the Afar Triangle. The Afar, Africa's most northern reach of the Great Rift Valley, is a geologist's dream because it is where three rifting systems intersect and buckle the landscape. It is the ideal place to study the process of continental drift. Maurice wanted to survey the area in hopes of piecing together its geological history. He called it "a geological and paleoanthropological paradise" and explained that it contained great expanses of sediments eroded by wind and water that were literally oozing fossils. Full of enthusiasm, he invited me to visit his lab the next day to see his photographs and maps of the place.

I stood in Maurice's lab electrified by his pictures of entire elephant skulls, pig jaws, antelope bones, crocodile skulls—complete specimens, not fragments like the Omo fossils. What impressed me even more was that the Afar pig and elephant fossils closely resembled the Omo samples that were in excess of 3 million years old. Maurice paced back and forth as he recounted stories of the Afar region, with its exotic people, its treasures, and its dangers. The Afar didn't like strangers, he explained. But once one got to know them they were wonderful. "You should go out there with me!" he said, exhaling thick, blue smoke and adjusting his heavy-rimmed glasses. I was floored; usually scientists weren't inclined to share information about a fossil-rich area, particularly if it was as promising as the Afar. It didn't appear to matter to him that I wasn't French: He just wanted to launch a full-fledged expedition to the region and, wonder of wonders, he wanted me to come along! A once-in-a-lifetime opportunity, for sure. I felt torn. I had to finish my dissertation, I had committed to another season in the Omo, and I had no money and no job lined up. But there was no way I could say no to Maurice; this was my chance to get in on the ground floor in a fossil paradise and perhaps find the hominid fossils of my dreams. "Okay," I said, "when do we leave?"

In March 1972 Maurice picked me up at Bole Airport in Addis Ababa. After quickly assembling food and equipment, we set off for a monthlong reconnaissance of the Afar. It was an endless expanse of badlands littered with tons of animal fossils exposed by erosion. "Terra incognita" Maurice called it, as each day we penetrated deeper and deeper into the uncharted territory. Escorting us was a spindly, gentle Afar man named Ali Axinum, who knew the region like the back of his hand and had served as Maurice's guide for several years. Without Ali we would have been lost, literally. The geol-

ogy of the Afar was overwhelming. Texan geologist Jon Kalb provided additional expertise. The Afar region held such an allure for
Jon that he moved his family to Ethiopia, where he had initially
worked for the Ministry of Mines. And rounding out our small expeditionary force was Yves Coppens, an ambitious French paleontologist I had met when he was directing the French team in the Omo.
Yves, like me, was especially motivated to find fossil hominids, and his
background in mammalian paleontology would be invaluable on
this first foray into the Afar region.

We arrived at Hadar in the afternoon. Maurice, who had
acquired his driving skills on treacherous French roads and packed
Paris streets, managed to brake just before he pitched our Land
Rover over a cliff. Hot, dusty, but mostly relieved when he stopped, I
climbed out and beheld a veritable El Dorado of paleoanthropology.
I stood on the edge of the vast expanse of eroded sediments confident that it was ideal: no vegetation, clear stratigraphy, and, most
important, fossils everywhere.

After our six-week reconnaissance we limped back to Addis
Ababa exhausted and suntanned but immensely satisfied with the
potential we had seen in the Afar for finding fossils. We formalized
our collaboration as the International Afar Research Expedition
(IARE) and decided that of all the places we had visited, Hadar was
the most promising. So in the fall of 1973, with a modest grant from
the National Science Foundation (NSF) and funds from the Centre
National de la Recherche Scientifique (CNRS), we set up camp at
Hadar. Day after day, bent over with my hands behind my back and
my eyes to the ground, I scoured the hillsides in search of that illu-
sive first hominid fossil. For weeks there was nothing. The heat,
effort, and constant backache began to sap my enthusiasm. But I
knew in my gut that it was just a matter of time and persistence. The
fossil hominids were there; they had to be.

Late in the afternoon of October 30, I saw what looked like a
hippo rib protruding from the ground. I gently tapped it with my

shoe, and the proximal end of a tibia (the top end of the shinbone) emerged from the loose soil. Crouching, I also spotted parts of a distal femur (the bottom end of the thighbone). When I placed them together, they formed a knee joint, missing only the kneecap. From the angle at which the shaft of the femur ascended I knew it was from an upright walking animal—the first hominid fossil ever discovered in the Afar region. I also knew that I now had important leverage to justify expanded funding for Hadar research.

While yielding valuable information about locomotion, the knee joint by itself held few clues about the species of its owner. When anatomist Raymond Dart discovered the Taung baby in 1924, he named his hominid *Australopithecus africanus,* the "southern ape of Africa." I surmised that the knee was *Australopithecus* but could not be certain whether it was the same species as Dart's or perhaps a new species. Despite the claims of some critics during my subsequent presentations that its small size indicated that it was nothing more than a monkey knee, I never veered from my original assertion that the knee belonged to a biped. The way the femur and tibia articulated with each other, making maximum contact, was a distinctly hominid trait.

With the fossil securely bundled in a small cardboard box labeled Afar Locality (A.L.) 128/129, I returned victorious to Case Western Reserve University in Cleveland where I had begun teaching, a big-game hunter bringing home his trophy from an African safari. I spent many humid summer nights completing my thesis, and late in August 1974 I successfully defended my daunting 444-page doctoral dissertation on chimp teeth. I had a little NSF money remaining and with some additional private funds I was able to enlist a few more students and colleagues who would accompany us back into the field. In early September we set off for Ethiopia and the next great adventure.

The IARE departed in a convoy of heavily laden Land Rovers, trailers, and a lorry from Addis Ababa. Even before camp had been fully set up, a few team members began eagerly searching the geological deposits within walking distance of Camp Hadar in hope of finding a hominid. One of the most motivated and keen-eyed was a young Ethiopian colleague from the Ministry of Culture, Alemayehu Asfaw, who was possessed by what I call "hominid fever." The outline of his figure could be seen roaming the hills until sunset, when it was too dark to survey. His perseverance eventually paid off when he found two jaws containing beautifully preserved teeth in the very deposits that others had searched near camp. Were they the same species as the owner of the knee joint? And was that species *A. africanus,* like Dart's Taung baby, or something new?

Friendly rivalry is at times a good thing because it motivates us all to work a little harder, a little longer, and take chances we otherwise might not take. Chief among my rivals was Richard Leakey, son of Louis and Mary, who had made some spectacular finds at Lake Turkana in Kenya, including a specimen known as 1470, then considered the earliest evidence of *Homo.* Tall and lanky with sun-bleached hair and boyish good looks, Richard exuded authority and confidence. We'd met several times, and the previous year I'd brought the fossil knee to Nairobi so that his team at the National Museums of Kenya could make a cast of it. He graciously allowed me to see their hominid fossil collection, including specimens that hadn't been published yet or even announced. Honored to be a member of the "inner circle" of scholars who were granted such access, I felt it was now my turn to return that favor, so I invited Richard; his wife, Meave; and his mother, Mary, to Hadar. A licensed pilot, he flew to Ethiopia in his own plane and landed at our hastily cleared airstrip, dubbed Hadar International Airport. In addition to Meave and Mary, Richard brought his brother-in-law, a paleontologist named John Harris. As they walked Hadar's astonishingly fossil-rich hillsides, I enjoyed watching their dumbstruck expressions, thinking that was how I must have looked to Maurice when I first came here.

Over dinner we debated the fossil identification of the new hominid jaws that Alemayehu had found. Richard favored classifying the larger one as part of a member of our own genus, *Homo*, whereas I thought it belonged to a male australopithecine. We weren't just arguing about the jaws, though. Richard believed that the 1470 skull was 3 million years old. But like many other experts in the field, I suspected that the geological dating was wrong because the animal fossils found along with it, particularly the pigs, were identical to those from the Omo that were dated at about 2 million years. Although we disagreed, I was thrilled to be exchanging ideas with some of the giants in the field. The stimulating conversation, which included students and other Hadar team members, left everyone excited but exhausted.

On Saturday we drove Richard and his group out to the airstrip, helped them load up, and waved good-bye as they took off for the flight back to Kenya. The following day I found Lucy.

During the 1974 field season at Hadar, Ethiopia's emperor, Haile Selassie, was overthrown by the Derg, a violent military junta who capitalized on Selassie's decreasing popularity following a poorly managed famine in the very region where we were working. I was deeply concerned about how this turn of events would impact our future plans to work at Hadar. Sure enough, when we returned in 1975, the political atmosphere had changed from calm to violence and fear. With increased funding from both the United States and France, the IARE team had grown to include more students and more specialists in paleontology and geology. I had always been concerned about medical issues in the field—dysentery, malaria, snake bites, appendicitis—so I was pleased when Mike Bush, a young, soft-spoken medical doctor with a passion for archaeology, expressed interest in joining my expedition. Not only was Mike a great guy to have in camp, as well as a competent physician, he also turned out to be a keen-eyed and avid surveyor. Late in the morning on November 2, Mike's dedication paid off. I was in the middle of a conversation with

David Brill, a *National Geographic* photographer, when Mike interrupted us to say that he had found what he thought might be some hominid teeth. David, a tall, skinny guy from Wisconsin, was excited because after hounding me for days, this might just be the photo op he had been pushing for—a hominid find. We followed Mike back to the spot.

"Right there," Mike said, pointing. I crouched, squinted, and was amazed to see two gray teeth embedded almost invisibly in gray stone. How had he managed to spot them? "Upper premolars," he added.

I grinned, looking up at David. "Here's the hominid I promised you."

David looked at the teeth, then at the sky. "You couldn't have discovered them when the lighting was better?"

I thought he was kidding, but he just shook his head and refused to take any photographs until the next day, when the sun would be lower on the horizon. We headed out to Mike's find first thing next morning, when the light was warmer and shadows gave the landscape a more three-dimensional feel. David began snapping away. I could tell he wasn't very impressed with the amorphous block of stone containing two little blobs of gray, but he was doing his best.

Hoping for a noteworthy discovery, Maurice had persuaded a French film crew to spend most of the field season with us. One of them had positioned herself upslope from us near a little bush, and suddenly she called down to me, *"Quels sont ceux-ci?"* She was holding up a couple of fossils, and as I approached her, much to my astonishment, I could see that one was a heel bone and the other a femur, both unmistakably hominid.

"Could this be another skeleton?" I yelled, and everyone scrambled up the steep hillside. It turned out to be a mother lode of hominid fossils: During the rest of that season and the next, location A.L. 333 furnished more than two hundred fossils, including a partial baby skull, portions of an articulated foot, some articulated hand bones, most of an adult male brain case, parts of faces, and numerous mandible fragments. Together these represented, in our best estimate, at least thirteen individuals—nine adults and four children or infants. From the title of my 1976 *National Geographic* article, the 333 collection was dubbed "the First Family."

Maurice speculated after careful inspection of the geological strata that the fossil-bearing layer was the result of a single geological event such as a flash flood; a snapshot of a group of hominids who lived, died, and were buried together. Unlike the usual discoveries of single specimens, the 333 assemblage offered unprecedented insight into the anatomy not only of a single species but of individuals who had actually known one another millions of years ago. Again, I marveled at the circumstances that had led to the discovery. Had I turned down Mike's request to join our expedition, the well-concealed teeth might not have been found, and A.L. 333 could have gone unnoticed.

By 1977 Ethiopia was under the ruthless leadership of Mengistu Haile Mariam, and it was too dangerous to conduct fieldwork. I was now the curator of physical anthropology at the Cleveland Museum of Natural History, and the major part of my job was to oversee the study of the unmatched hominid harvest from Hadar. The process involved cleaning the stone matrix from each piece of bone in order to obtain a clear picture of its surface anatomy, taking color and black-and-white record photos, molding and casting each specimen, as well as describing each fragment in meticulous detail for publication. All of this had to be accomplished by 1980, when the fossils would be returned to the National Museum of Ethiopia for safekeeping.

I assembled a team—consulting experts, researchers, students, and colleagues—to systematically undertake a comprehensive study of the Hadar collection. Mike Bush used his expertise in anatomy and his interest in hands to catalog the hand bones; Owen Lovejoy, a leading expert on hominid bipedalism and postcranial anatomy at Kent State University, was assigned to the limbs and the pelvis; William Kimbel, Lovejoy's graduate student, worked on the cranial remains. I recruited Tim White of Berkeley to document the mandibles. Tim, a promising young paleoanthropologist with a curmudgeonly streak, had worked with Mary Leakey at Laetoli, in Tanzania, where similar fossils had been found. We had met in 1975 when I brought the First Family to the National Museums of Kenya, where he was working for the Leakeys, and we struck up a friendship. Having parted ways with Richard's expedition and then Mary's, he joined mine. Meanwhile, ever the dental expert, I worked on the teeth. The final analysis was eventually published in a 352-page single issue of the *American Journal of Physical Anthropology.*

While we were working on detailed descriptions of the bones in early June 1977, I was unexpectedly invited to present an evaluation of the Hadar hominids at a Nobel Symposium in Sweden. Convened by the Royal Swedish Academy of Sciences to address the "Current Argument on Early Man," the symposium commemorated the two hundredth anniversary of the death of the Swedish botanist Carolus Linnaeus, whose taxonomy (the practice of describing, classifying, and assigning Latinized binomial names to all biological organisms, living or extinct, for placement in a hierarchical system) is still the basis of modern classification. I could think of no occasion more fitting than a Nobel Symposium to present our results. My parents had emigrated from Sweden in the early 1900s, and I felt honored to deliver a paper in the country of my ancestors. I knew that some of paleoanthropology's most accomplished scholars would be in the audience listening carefully for speculative claims or factual errors. My presentation had to be meticulously researched, and I was grateful for colleagues like Tim and Owen and others who agreed to review draft after draft, sometimes offering brutal criticism.

On May 21, 1978, a clear, cool Sunday evening in Stockholm, I attended a gala reception and dinner hosted by the Royal Swedish

Academy of Sciences. Early the next morning we set off by bus for the town of Karlskoga, some 240 kilometers west of Stockholm, home of Bofors, the major sponsor of the symposium. Alfred Nobel had owned the company in the late nineteenth century when it became important in the chemical industry and the manufacture of cannons. Unfortunately, I was too preoccupied with my upcoming presentation to appreciate the passing countryside.

Richard Leakey opened the conference with a relatively short talk on his work at Lake Turkana in Kenya, describing the many hominid discoveries found there. Then it was my turn to walk to the podium. With my stomach in knots, I presented my paper, "Early African Hominid Phylogenesis: A Re-evaluation," describing the Hadar hominid collection and delivering evidence that they represented the same species as fossils found at Laetoli, in Tanzania, by Richard's mother, Mary. I made my case that all of the fossils bore more primitive anatomical features than any other known hominid species. And then I announced that these hominids belonged to a new species: *Australopithecus afarensis*. As I listened to my words reverberating through the small auditorium, I thought about the sixteen-year-old who dreamed of finding the missing link: He was now standing, dressed in a three-piece suit, before the most distinguished audience on the planet, publicly pronouncing Lucy's scientific name for the first time. It was a provocative thing to do. Many researchers thought that the fossils were simply east African versions of *A. africanus*.

From the sound of people shuffling in their seats and a few audible aspirations I could sense the immediate skepticism. My mouth went dry and I took a sip of water. In my next slide I showed a redrawing of the human family tree and elaborated on the major implications of the new species for understanding the shape of that tree. I was fully aware that most of the assembled favored placing Dart's Taung baby, *A. africanus*, at the base of the tree, the common ancestor to all later hominids, including modern humans. But I boldly maintained that it was now time to relegate *A. africanus* to an extinct side branch of human evolution, and to position *A. afarensis* as the trunk of the tree, the last common ancestor to all post-3-million-year-old hominids. Dead silence. I took another sip of water, the lights came on, and Carl Gustaf Bernhard, the moderator, called for questions. Not a single hand went up. The silence was broken when the moderator suggested that we break for high tea.

During the remaining days of the symposium I continued to argue the main points of my paper, but most of the guests were unconvinced. How could I be sure that all the Hadar hominids were of a single species, they wondered. Based upon what evidence did I put Hadar and Laetoli fossils in the same species? I knew that my colleagues and I faced months of work, possibly years, before our conclusions were accepted and Lucy could assume her proper position on the human family tree.

Unfinished Business

January 3, 1980, was a frigid morning in Cleveland, and it was with apprehension that I transported my treasures to the airport. I checked my luggage, but kept Lucy and the other hominid remains in my carry-ons. In those days, it was common practice for paleontologists to transport hominid fossils out of their country of origin for study and later return them. In our case, we had signed an agreement in 1975 stating that after five years we would return the Hadar specimens to the National Museum of Ethiopia for permanent storage. During that time, we cleaned, photographed, molded, studied, and described the fossils. Today, however, such remains typically stay in the country where they were found. I was quiet during the trip, moody and distracted. I've never been afraid of flying, but that day I couldn't stop imagining the grim headline: "Flight Crashes, Famous Fossil Disappears into Ocean."

In Paris I met up with Maurice Taieb, who would accompany me to Ethiopia. A wonderful conversationalist, he distracted me with his many colorful stories, opinions, and speculations. The ten-hour flight passed quickly, and before I knew it, the time had come and we were handing the Hadar hominids to Mamo Tessema, the museum's director.

"We feel such pride that she is ours," he said as he opened the

case and admired Lucy. "She belongs to us. We are the cr\
humanity; she is proof."

"Bien sûr," agreed Maurice. "Absolutely."

I sighed. Of course she belonged here, there was no question about it. But leaving her felt like putting my child up for adoption: There was paperwork to be signed, hands to be shaken, and I was left wondering forlornly when I would see her again.

Other staff members and researchers doing work came over to see her. I recognized a lot of them, and they greeted me with half glances before turning their attention to her. Once again I watched her cast her spell and consoled myself with the thought that she was only the beginning of what the Afar region would offer up. After returning to Cleveland in the middle of a bitterly cold February, I sat in my office and stared woefully into the empty safe. Having lived with Lucy for five years, I was experiencing genuine separation anxiety. Fortunately, there was a lot to keep me busy. We were polishing up our reports on the Hadar hominids to get them ready for publication in a number of scientific journals. And I was happily and heavily occupied with completing my first book, *Lucy: The Beginnings of Humankind.* My coauthor, Maitland Edey, made periodic visits to Cleveland, and I would occasionally travel to see him in Martha's Vineyard, where he had a second home.

About midway through 1980, while Tim White and I were doing a comprehensive study of *Australopithecus* teeth and jaws in South Africa, it dawned on me that not only was Lucy's time in Cleveland over, but mine was, too. I had worked in Cleveland for eight years, first as a young assistant professor at Case Western Reserve University and then as the curator of physical anthropology at the Cleveland Museum of Natural History. I had built a wonderful laboratory at the Cleveland Museum and done much to put it on the map. I suppose I could have remained there for my entire life, but as usual I was getting restless and wanted a new challenge.

One of the dominant paleoanthropology programs in the

country was at the University of California–Berkeley. I visited Clark Howell there several times and hoped to someday live in the Bay Area. Encouraged by discussions with Tim and Clark, as well as financial support from notables like Gordon Getty, I made the decision to leave Cleveland and become the founding director of the Institute of Human Origins (IHO)—originally known as the International Institute for the Study of Human Origins—in Berkeley.

It was not easy for me to leave Cleveland, where I had so many friends and a nice home. I will always have pleasant memories of my time there. The Cleveland Museum of Natural History had played a pivotal role in my professional development, and the staff had been unselfishly supportive of all my research. My old lab still functions and is currently under the able directorship of a young Ethiopian paleoanthropologist, Yohannes Haile-Selassie.

By late 1981 I had relocated to Berkeley and faced the challenges of setting up a nonprofit research institute. I established a board of directors and began fund-raising efforts to realize my dream of having a human evolution think tank solely focused on paleoanthropology. I didn't let my institute duties eclipse my hopes of a return to Hadar, though, and I kept in contact with the Ethiopian Ministry of Culture. My close colleagues at Berkeley—J. Desmond Clark, a doyen of African archaeology, and Tim White—began fieldwork in the Middle Awash region of Ethiopia in the fall of 1981. And their success encouraged me to apply for a permit to work at Hadar in the autumn of 1982.

Constrained by my institute responsibilities, I had to remain in Berkeley after I sent the Hadar team ahead to Addis Ababa to make preparations for fieldwork. Gerry Eck, whom I had met in the Omo, was key to these preparations. Gerry was one of my closest and oldest friends, but I wasn't happy to hear from him when he called at two o'clock one morning from Addis. Groggy, I asked, "Everything all right?"

"You need to be here now," he said. Dread swamped me.

"What's going on?"

"They're talking about rescinding everyone's permits!"

"What? Why?"

"The official reason is that the Ministry of Culture has decided it's time to overhaul their regulations governing archaeological research in Ethiopia."

"But . . ."

"Apparently scholars at Addis Ababa University have been complaining that not enough of them are being included in field research. They say it's all foreigners."

"Meaning us?"

"Meaning us. They're pressuring the ministry to overhaul the whole system—start favoring native researchers, train the locals, and improve the university and the National Museum."

I was at a loss for words. The reasons were more or less valid, but it didn't seem fair to those of us who had made such stunning discoveries and brought recognition to Ethiopia. The rules and regulations guiding archaeological investigations were antiquated and certainly required updating, but why choose this moment to make such a devastating decision and declare a temporary moratorium on fieldwork?

Four days later, on September 19, 1982, I arrived in Ethiopia. Gerry picked me up, looking grim. "The Berkeley and Hadar teams have completed preparations, and now they're just sitting around, waiting to be told they can leave for the field. About two dozen of them, raring to go."

"This could be bad," I said.

"This is bad, Don. The Ministry of Culture is still waiting for a new minister to be appointed so they're not in a strong position to respond and negotiate all the accusations."

"You think they can actually shut us down?"

"I think they can, and they will."

Sure enough, less than a month later the Ethiopian government ordered cessation of all archaeological work in Ethiopia. Like it or not, we all had to pack up and go home.

Most of us who have spent major portions of our lives living in a tent agree that the best part of paleoanthropology is the fieldwork: solving the problems of geology, archaeology, paleontology, and experiencing the thrill of hominid discovery. The field, with all of its challenges, is the place to be. It's where the alpha evidence resides, and where new discoveries can make careers and alter our understanding of human origins. As recently as the late 1970s, the entire storehouse of pre-3-million-year-old hominid fragments would fit into the palm of your hand. This is why work in the field is never complete; it can never be considered "done." Louis Leakey first excavated Olduvai Gorge in 1931, and seventy-seven years later expeditions there continue to make important discoveries.

The situation was no different at Hadar. Not only were there more discoveries to be made, but key questions about the biology and adaptations of the Hadar hominids and their geological and paleoenvironmental contexts could only be solved, augmented, consolidated, and solidified by going back into the field. Everyone felt helpless and frustrated. We were like kids benched game after game: All we could do was wait until we got the call.

There was plenty to do, however. I founded the Institute of Human Origins in 1981 and built it into a nonprofit research facility in Berkeley. And our analyses of the *A. afarensis* fossils we had collected thus far at Hadar led to the publication of numerous scholarly articles. We also spent a few field seasons exploring Olduvai Gorge.

We'd been told we could reapply for permits within a year or so, but it was not until 1989 that the Ministry of Culture, now called the Ministry of Culture and Sports Affairs, lifted the ban and I received a letter from Tadessa Terfa, the head of the Centre for Research and Conservation of Cultural Heritage, formally inviting researchers back into the field. I immediately arranged a meeting in the IHO library with Bill Kimbel and Bob Walter. Bill, a highly respected expert on *A. afarensis,* served as the science director at IHO. Bob was a geologist who had done his Ph.D. on the volcanic history of the Hadar Formation under Jim Aronson at Case Western Reserve University, and then later joined the Berkeley Geochronology Center, the division of IHO that was focused on calibrating the earth's history.

"We're in this together, you guys," I said, "and I think we should be codirectors of the expedition."

There were prompt nods of agreement, and Bill, always pragmatic, didn't waste any time getting down to business. "First thing we have to do is get funding," he said. "We secured commitments that will allow us to purchase two new Toyota Land Cruisers, but we need money for airfares, per diems, and so on. I'll get to work right away crafting a National Science Foundation proposal."

"Okay, and I'll approach National Geographic." I flipped open my notebook and uncapped a pen. "Let's outline our objectives and strategies."

Bob noted that we needed to get a more concise date for the fossils. The only way to definitively determine the age was to collect samples of the surrounding volcanic horizons and analyze them. I wrote: *Collecting and analyzing volcanic ash samples is a top priority!!!* I wasn't totally current on the latest technology, but I knew all too well that radiocarbon dating—which is what everyone thinks of when they think of fossil dating—was only useful for organic material, such as carbon, wood, shell, or bone going back about 50,000 years. In the specimens at Hadar, believed to be over 3 million years old, minerals had long since replaced their organic material. Technically, fossils are stone, not bone, because organic material is replaced molecule by molecule with minerals that are in the water. The only way to determine their age is to perform potassium-argon dating analysis on the surrounding geological strata, specifically the layers containing volcanic ash or lava. We had one such potassium-argon date from a layer above the one that Lucy came from, and we knew as a result that Lucy had to be more than 2.8 million years old. And the pigs and elephants at Hadar corresponded to the pigs and elephants in strata well dated by potassium-argon in the Omo region. So we knew Lucy was at least 3 million years old, but we didn't know how much older.

Potassium-argon dating made its paleoanthropological debut in 1959 at Olduvai Gorge, when it was applied to a volcanic ash horizon just above the layer that yielded Mary Leakey's "Nutcracker Man" cranium. The date of 1.8 million years for the skull astonished many paleoanthropologists, including Louis Leakey, who believed it to be younger. As the first fossil to have a reliable age estimate that old, the find ignited human origins research in eastern Africa along the Great Rift Valley. Previously researchers had focused on South Africa in the search for hominid remains. But in South Africa, the fossils are found in caves, whose contents tend to become jumbled during deposition. In eastern Africa, by contrast, the fossils occur in well-stratified sequences that go from older to younger. Like moths drawn to a flame, anthropologists flocked to eastern Africa with a fervor akin to that of the Gold Rush, but in this case it was the Hominid Rush.

Potassium-argon dating is really rather straightforward. We know that volcanic rocks are rich in potassium-40 (^{40}K), an unstable isotope of potassium. This ^{40}K decays over time at a predictable rate, and one of the by-products is argon-40 (^{40}Ar), a gas that gets trapped in the crystal lattice structure of minerals. So the number of ^{40}Ar atoms in a rock sample is a function of time—the older the rock, the more ^{40}Ar atoms. Unfortunately, eons of weathering has altered many of Hadar's layers of volcanic rock and robbed them of all their ^{40}Ar, or they have become contaminated with older volcanic materials. Therein lies the trouble with potassium-argon dating: Data can vary enormously depending on external, unavoidable, and often unknown factors, and a rock dated to 2 million years might have a plus or minus error anywhere from 40,000 years to 250,000 years. We were eager to rein in the numbers to something more precise.

"I know Derek York at the University of Toronto has made some major advances in his geochronology lab," I said. "Has he got anything that can be applied in the Hadar geology?" Bob explained that York had pioneered a method called single-crystal laser-fusion (SCLF) dating—the absolute latest technology. The most common isotope of potassium is ^{39}K, which is very stable. And we find it everywhere, including in volcanic rocks. When you analyze a volcanic rock sample, utilizing the traditional potassium-argon method, the rock has to be cut in half. One half yields the percentage of potassium and the other half is analyzed for the number of argon atoms. The basic premise here is that the chemical makeup of both halves is identical, but rocks are notoriously heterogeneous, which can lead to age estimates that have rather large plus or minus errors. Derek's SCLF method promised to produce more precise estimates of when that volcanic rock actually formed.

"So tell me about the laser fusion," I said.

"Okay. I know you appreciate elegance, so listen up." Bob had a gleeful look in his eye, and I knew he couldn't wait to dazzle me with a bit of science. I leaned forward, all ears. "What if we could measure the amount of potassium-39 in the exact same sample we were analyzing for argon-40?" he said.

"You'd get a more exact estimate of a sample's age," I replied. "But how can you do that?"

Potassium-39 is stable under most conditions, Bob continued. But if it's bombarded with neutrons, it converts into argon-39. By inference, then, when the mass spectrometer measures the argon in the purified gas, the amount of argon-39 represents the amount of potassium in the sample. "Brilliant, eh?" He grinned.

"Not bad!" I remarked. "But what about the laser?"

Bob explained that the next step was to isolate crystals of feldspar—a mineral known to be high in potassium—from the volcanic ash, irradiate them, then put them into a vacuum chamber, where a laser beam would melt them and a mass spectrometer would analyze the resulting gases.

"That's way more precise," I observed.

He nodded. "Even if we only find a limited number of feldspar crystals, we might still be able to date the volcanic horizons."

"I'll write the proposal to reflect the importance of gathering volcanic samples," Bill said. "Establishing a more exact date for the Hadar fossils will affect all future discoveries in that region. That's a huge justification for funding.

"But we also need to find more hominids," he continued, tapping the notebook impatiently as I wrote *Find more hominids!!!* as if otherwise I'd forget to look for them. "There are still some issues about the systematics and paleobiology of *afarensis* that are under debate."

Bill was referring to the nagging issue of whether all the specimens we assigned to *afarensis* represented one species or two. It was unbelievable to us, but some of our colleagues thought there might be two species. They didn't like the extreme morphological and size variations in the Hadar samples. We were pretty sure that the differences in size were related to sex, with the males being much bigger than the females, and that the variation in anatomy was a feature of Lucy's species. It was also possible that we were dealing with evolutionary change over time within a single species. Our conclusion: We needed to find more hominids, especially from the younger levels at Hadar, to firm up our case for a single species.

I sighed. Find more hominids. No problem! Why didn't I just discover a cure for cancer and solve the world hunger problem, too? When we named *Australopithecus afarensis*, we thought we had an unassailable case for the species. We had followed the well-established practice of comparing levels of variation in tooth size, mandible shape, body size, and so on in the Hadar collection with levels of variation for the same anatomical trait in collections of modern apes. The variation within the Hadar sample did not exceed that within modern ape species, our very close evolutionary relatives, so we concluded that there was only a single species at Hadar. Much of the size variation we attributed to sexual dimorphism. Then we compared, anatomical trait by anatomical trait, the Hadar material to its closest fossil relative, *A. africanus*. The Hadar sample was demonstrably much more primitive than *A. africanus* in so many dental and cranial features that it just had to be a different species, *A. afarensis*.

"Jeez, Bill," I exclaimed. "How many more specimens do we need to justify our single-species argument?"

Bill shrugged. "Hopefully a lot of the issues of taxonomy and sexual dimorphism will be resolved when we find some relatively intact skulls of *afarensis*. That's what we need more than almost anything—skulls."

Maybe we can find some of their diaries and newspapers, too, I thought sourly. In the late 1970s, Bill and Tim White had assembled a "plaster hypothesis" of an *A. afarensis* skull using 107 fragments of several individuals at Hadar, mostly from A.L. 333. That's all we had, but it wasn't enough to convince some scholars. Someone even suggested that the face belonged to one species and the brain case to another.

Either way, Bill went on, we could use more fossils to resolve the issue of how *A. afarensis* got around. We were certain that *A. afarensis* walked on two legs, but some of our colleagues argued that aspects of the functional anatomy of the shoulder, elbow, hip, knee, ankle, and foot indicated a significant degree of apelike arboreal climbing.

"We need more locomotion-specific fossils," Bill said. "Add it to the list."

"Intact skulls and complete skeleton," I sighed. "Anything else?"

"No, that'll do." Bill leaned back in his chair. "I've been looking at our aerial photos and discussing the geology with Bob," he said.

The two had determined that there were large expanses of unexplored sediments in our permit area that sampled the upper reaches of the Hadar Formation, which were chronologically closer to 3 million years, if that single potassium-argon date we had was correct. Hominids from there would help close the evolutionary gap between our *A. afarensis* specimens and post-3-million-year-old hominids. They might also provide a clearer picture of the evolutionary relationship of *A. afarensis* to later species of *Australopithecus*, such as *A. africanus* and *A. aethiopicus*. We fell silent, each lost in his thoughts. No matter how daunting the challenges that lay ahead of us were, the important thing was that the ban on fieldwork by foreign researchers in Ethiopia had been lifted.

We submitted our proposals to the National Science Foundation and the National Geographic Society, and sent our application for fieldwork to the Ethiopian Ministry of Culture and Sports Affairs. Our initial draft had been ambitious to the point of being overwhelming, so we made some prudent adjustments. Knowing that the National Museum of Ethiopia was bulging at the seams with fossil fauna collected from Hadar in the 1970s, we stated our intentions to limit our collection of nonhominid fossils. The only exception would be especially complete specimens of underrepresented, rare, or new species. We would also keep our eyes open for certain animals like pigs, which can be useful in making age correlations with other sites that have been well dated (this technique is called biostratigraphy). And we decided not to undertake excavation unless absolutely necessary—at a promising new hominid locality, for example. This meant that we would visit A.L. 333, the First Family site, but only to assess the status of erosion and formulate a strategy for excavation at a later time. We had a lot left to do at A.L. 333, and it would have taken too much manpower for this trip.

Our research plate was deliciously full, our spirits were high, and we all knew the prospects for reward were as promising as ever. Hadar was waiting to give up more of its secrets to us—we just needed to get there. When a reporter heard about our plans and asked me what I expected to find, I answered simply, "The unexpected!"

CHAPTER 3

Rocky Beginnings

A s Alitalia 802 from Rome made a wide turn around the capital city of Addis Ababa on September 28, 1990, I spotted Bole Airport and anticipated the landing. I could see the blue haze of smoke hovering over the city from the many charcoal fires that kept people warm during the cold nights at an altitude of 7,726 feet. It had been fourteen years since my last major expedition to Hadar.

Since 1982 the Ministry of Culture and Sports Affairs had imposed a moratorium on paleoanthropological work, and the Ethiopian Revolution was also cause for caution. It was led by Mengistu Haile Mariam, a ruthless, brutal, and coldly calculating Marxist who was rumored to have personally smothered Emperor Haile Selassie to death. In 1977 Mengistu launched the Red Terror to crush any opposition to his Provisional Military Administration Council and brought about the deaths of hundreds of thousands, including many of Ethiopia's best-educated and most affluent citizens.

Now firmly within the steel grip of Mengistu's regime, the People's Democratic Republic of Ethiopia (PDRE) had reached a state of semistability by 1990, which led the Ministry of Culture and Sports Affairs to lift the ban on fieldwork and issue invitations to research groups like mine to again undertake paleoanthropological investigations. I had patiently waited for this day, and as the plane

landed I let out a deep sigh. It had been a difficult and long banish-
ment for paleoanthropologists like myself who had nightmares
about spectacular fossils eroding to the surface, only to disappear
again into the ever-changing geography. As I stepped into the blinding African sun I was so ecstatic to be back that I nearly knelt down
to kiss the tarmac.

With my camera bag slung over my shoulder and another carry-
on containing irreplaceable items necessary for the field, I made my
way toward the airlines transfer bus. Surrounded by Ethiopians
dressed in colorful cotton shawls called *gabis*, I enjoyed the familiar
sound of soft conversation in Amharic. When we arrived at the ter-
minal, we flooded out of the bus, jockeying for a slot in one of the
slow-moving immigration lines. Finally stepping up to the kiosk, I
slipped my passport and arrival form through a small opening into
the waiting hand of an immigration officer. After a thorough
scrutiny of my passport and visa particulars, I heard the welcome
thud of the entry stamp, and with a somewhat guarded look from
the officer I was waved on to the baggage claim area.

While waiting for my luggage, I scanned the crowd, inexplicably
hoping to see someone I recognized. Right away I picked up on a
different vibe: an overwhelming sense that these people were sad.
Ethiopians are normally cheerful, upbeat, and welcoming, but that
day all I saw were long faces and dreary expressions. Fifteen years
under Mengistu's rule had deadened their spirits.

The sound of a diesel engine outside heralded the arrival of a
baggage train stacked high with our luggage. The squeaky, dilapidated carousel lurched into motion, and through a small, square
opening covered with hanging straps of rubber, suitcases began to
trickle out. As I loaded my two bags onto a cart, I remembered how
much I dreaded going through customs, where I could already see
half a dozen yanked-open suitcases with contents spilling out—
clothing, kitchen implements, radios, tape cassette players, unworn
shoes, boxes of laundry soap—items that were remarkably expensive

and almost impossible to find in Ethiopia. The customs officials were taking their time, scrutinizing everyone's baggage, looking for anything they could charge an import duty on. So when a stony-faced official caught my attention and waved me over, I thought, Uh-oh, here we go.

I had passed through immigration and customs in Addis more times than I could remember, but I still got butterflies in my stomach at the thought of catching an official on a bad day and having my luggage thoroughly searched, as I was forced to justify every single item I had so carefully packed at home. After my long absence, I was even more anxious. I wondered if the process had become more rigid.

I slowly pushed my cart toward the tall, thin customs officer and tried to muster a little smile. He stood waiting with his hand outstretched. In a cold, surly voice, he demanded, "Passport and currency form!"

"*Tenastiling,*" I greeted him timidly, handing over the papers. He looked startled, then his expression softened.

"Oh, you have been to Ethiopia before?"

I nodded. "Yes, have you ever heard of Dinkinesh?"

His eyes grew wide, and to my delight, he said, "Lucy?"

"Yes!" Eager to please, I tapped my chest. "I found her, sixteen years ago!"

He grinned. "Welcome back! Ethiopia should be proud to have you here!" With a piece of chalk, he approved each bag, and said, "You are welcome!"

"*Amasaganalu,*" I replied. Thank you.

Relieved, I moved on. An older man dressed in khaki overalls gently came between my luggage cart and me. He smiled and said, "*Denano?*"

"*Denano,*" I repeated. Okay. The skillful porter threaded our way through a thicket of people to the narrow passageway leading to the parking area. I searched the crowd until I spotted the friendly and familiar faces of Bill Kimbel and Bob Walter. In that instant the last of my anxiety vanished, giving way to elation. I was back in Ethiopia!

I pointed. "My friends," I told the porter, and he nodded.

I hurried over to greet them. Walking to the parking lot, we swapped small talk and they inquired about my flight. Then Bill, like a typical, proud new car owner, swung open the rear door of our sparkling, white Toyota Land Cruiser and said, "Here she is!"

"Well," I sighed, "I guess it was only a matter of time before our Land Rovers would give up the ghost."

"Don, those Rovers were twenty-five years old."

"Still, though. Lots of good memories." In those past days the masculine-looking Land Rover was de rigueur: It handled like a tank, broke down more often than not, and was about as comfortable a ride as going over Niagara Falls in a barrel. Like a new, more powerful, better-adapted animal entering the landscape, Toyota Land Cruisers had invaded Africa, slowly driving the Land Rovers into extinction. Abandoned, they accumulated gracelessly in Land Rover graveyards.

The porter loaded my bags into the Land Cruiser. Bob and I got into the front seat, and with Bill at the wheel, we left the parking lot and merged onto Bole Road, the jammed artery leading into the city center. If you don't know how to use the horn, you get absolutely nowhere in Addis—the roads are choked with herds of sheep, goats, and donkeys weighed down with teff, a bitter grain used to make *njera,* a flat, pancakelike bread that is the cornerstone of the Ethiopian diet. Accompanying them is a seemingly endless procession of Ethiopians walking into town. Sharing the road with the small blue Fiat taxis, the kamikazes of Addis, is a little like riding bumper cars at a circus. Their daredevil maneuvers cause pedestrians and animals to literally jump out of the way to escape being hit. Bill, a fast but excellent driver, almost anticipated the moves of people, taxis, and animals ahead, a skill that had come from years of experience navigating the challenging and now largely potholed and neglected roads in Addis.

Although we had stayed in phone contact during the several weeks that Bill and Bob had been busy in Addis, I was eager to hear the latest news.

Bill swerved to avoid a barefoot woman overloaded with wood. "Actually everything is proceeding without a hitch . . . it's sort of scary."

"Our permit went through?"

"Uh-huh," Bill replied.

"No delays?"

"Nope."

"The Ministry of Culture got security clearance?"

"Yup. We're good."

"That is scary."

"We could still have trouble getting through police and army checkpoints," Bob offered, "but it's not likely. Registration of the vehicles took some time—the usual bureaucracy. But we got our stickers."

"Did you get supplies? Drums for water? Tent flies, chairs, tables?"

"Yeah." He consulted a list in his shirt pocket. "Plus kitchen equipment, tools, spare parts, medical supplies, canteens, excavation equipment, aerial photographs."

"So we're basically ready to leave for the field?" I asked.

"Basically."

"Outstanding."

I sat forward, practically hanging out the window, and took in the sights, sounds, and smells of exotic Addis Ababa. We drove through Revolution Square where Mengistu held his rallies and threw bottles of red paint into the crowd symbolizing the blood of his enemies, and I saw large slogans of the PDRE prominently displayed and accompanied by a hammer and sickle. We passed the old imperial palace, where Haile Selassie once lived, continued past the Addis Ababa Hilton, and at the top of a steep hill turned left onto the road leading to the National Museum of Ethiopia, where Lucy and the avalanche of fossils that followed were carefully stored.

Just as we drove into the museum entrance, a few blocks past Addis Ababa University, I looked at Bill and said, "Why do you think things are going so well?"

Slowing down to greet the police at the museum entrance, Bill downshifted. "Well, if it makes you feel better, there is one thing that could be a potential problem . . ."

"Oh?" I could feel my stomach clench. I knew it was too good to be true! "What is it?"

"I'll tell you later."

Bill coasted to a stop. He leaned out his window and greeted the police guarding the museum; they waved us on without asking to search the car.

I was left to speculate about what he meant, and my anxiety returned. Then I heard someone shout my name, and I saw Ale-mayehu Asfaw, the fossil hunter who had found the two ancient jaw-bones near Camp Hadar just prior to the discovery of Lucy. I climbed out of the Land Cruiser and hurried over to shake his hand and embrace him, pressing my right shoulder and then my left to his own in typical Ethiopian fashion. "Great to see you!" I said, conceal-ing my surprise at his appearance. His youthful look had been replaced by a serious expression, and his face looked lean and weathered.

As other friends came out to greet me I felt happy and more at home; everything felt familiar, as if I had never left. As I caught up on everyone's news, I tried to ignore the nagging worry about Bill's announcement. The museum was only a couple days' drive from the Lucy site, but I knew that a delay could turn into days, weeks, months, or even years. Inside there was a beehive of activity as other expeditions prepared to go to the field, including J. Desmond

Clark's group from the University of California–Berkeley, who had secured a permit to work in an area known as the Middle Awash, south of Hadar. Anyone else surveying this hectic scene would have come away with a sense that everything was normal, but I had worked long enough in Ethiopia to sense that the ground was shifting beneath my feet, and I would not relax until I was snug in my sleeping bag in my little tent overlooking the Awash River that ran through Hadar.

I was driven to the Ethiopia Hotel to get settled into room 309 and take a much-deserved nap after the draining flight from Europe. I unpacked and stretched out on the bed, but my mind simply wouldn't shut down, and all I could think about was the potential problem Bill had mentioned. Eventually I dozed off but was soon awakened by the sound of the phone ringing.

"Hello?" Crackling. "Hello?" I said again. Finally, I heard Bill's voice. "We're downstairs, do you want to meet us for dinner?"

"Be right there." I hung up and sat on the edge of the bed trying to get acclimated. I was still a little groggy. It was dark outside, and the city felt foreboding and mysterious—different than it used to. I looked at my watch: 8:00, which meant it was 9:00 a.m. in California. I was hungry, so I got up, washed my face, brushed my teeth, and combed my hair.

Instead of waiting for the painfully slow elevator, I took the stairs down to the ground floor. Men in dark suits scurried about, whispering to each other. I took the first right and entered the restaurant. Like the rest of the hotel it was dimly lit and sparsely populated, and I was amused to hear the same music I'd heard playing there in 1982; the tapes were even more worn, the songs less clear and wobbly as if coming from a warped record.

Immediately I was approached by the same headwaiter who had served me in 1974. He wore the same black quasi tuxedo that was now so shiny you could practically see your reflection in it. "Dr. Johanson!" he greeted me warmly. I shook his hand.

"You remember me?" I asked, pleased.

"Of course! We do not get many *ferenges* here."

Foreigner, yes, that was how I felt; even though every street was familiar, the atmosphere was different.

He led me over to a corner table where I saw Bill and Bob talking with Yoel Rak, our colleague from Israel. Deep in conversation, they sipped Coke and Ambo, a natural, fizzy spring water popular throughout Ethiopia. I sat down and decided to have spaghetti, the best bet on the menu, and a bottle of Bati, the local brew. With few customers in the restaurant, the two waiters had time to be extra attentive, bringing rolls and butter and Ambo, and taking my order.

"Well," I said, "how bad is it?"

"It isn't exactly 1982 all over again," Bill answered, "but things could get sticky, depending on what you decide to do."

"Me? Why me? I just got here. I thought everything was fine?"

"I've had a few, er, discussions with Berhane," Bill said, with a glance at Bob.

"About what?" Berhane Asfaw, who had done his Ph.D. at UC Berkeley, was the director of the National Museum. He was well known to all of us since he had worked in our casting lab at the Institute of Human Origins making plaster copies of Lucy for sale and even participated in my expeditions to Olduvai Gorge in the mid-1980s when work was not possible in Ethiopia. "I saw him at the museum today and he didn't say anything to me about potential problems."

Bill, looking more and more serious, replied, "Berhane tells me that there are some Ethiopian TV people who want to interview you. They want to ask you about some of the things you wrote in your book."

"What do you mean? What kinds of things?"

"Well, some of what you said about Ethiopia and Ethiopians wasn't very flattering. Berhane says these questions won't be nice.

They're going to try to put you on the spot. You have to get out of town."

"Are you serious? You're not serious." I'd heard before that some Ethiopians were unhappy with the book, though no one had ever told me why. But Berhane and I had never been very close friends. Was he just trying to make things difficult for me? It didn't matter. I couldn't take that chance.

Bill nodded somberly. "Berhane said if the TV people get to you, it will not only be embarrassing to you, but—and these are his words—your conduct may jeopardize the renewal of paleoanthropological work here."

"Shit!" I said, stunned and intimidated. "So what do we do? I'm willing to say or do whatever I have to to keep the expeditions from being closed down."

"You have to get out of town."

"What?"

"You have to disappear. Make yourself unavailable for interviews. Go into isolation." Bill stopped talking and leaned back as the waiter arrived with my beer. I thanked him as casually as I could, and as soon as he left I said, "I don't believe this!"

"It's the only way. Disappear and stay out of sight until the day, the hour, we leave to go into the field. Do you understand?"

"But I . . . I just got here!" I took a gulp of beer. "I don't feel like getting back on a plane! I suppose I could go down to Nairobi . . . but gimme a break! This is ridiculous!"

Bill looked around the room, leaned forward again, and said in a soft voice, "Berhane and Tim White think you can sneak out of Addis on Sunday and go to Nazret, hide out in a hotel, and be picked up when we leave for the Afar. What do you think?"

"What do I think? I think I'm suddenly starring in a film noir."

"But you'll do it?"

"There's no other way? What if I talk to them?"

"Don, no. That won't work. You don't realize how bad it's gotten here."

"We wouldn't suggest this if there was any other way to take care of it," Bob added.

I sighed. "Okay, I'll go."

On Saturday, while everyone else was at the museum packing, having vehicles serviced, or buying food supplies, I stayed in my room, calling hotels in Nazret, a town southeast of Addis, on the way to Hadar, hoping to make a reservation. But it was Mawlid an-Nabi, a holiday that celebrates Muhammad's birthday, and all the major hotels in Nazret were booked solid. The best I could do was to reserve a room at the Adama Ras Hotel for Tuesday.

Dejected and marooned, I stared out my window. It was the end of the rainy season in Addis, and the dark, gloomy sky reflected the mood of the city. This revolution, like so many that had come before it, was not kind to the poor and homeless. Barefoot children wearing T-shirts and shorts shivered in the evenings, when temperatures dropped down into the fifties, their faces numb as they asked passersby for a handout. Where would they pass the night? Probably they found sparse shelter in the bushes lining the hotel grounds.

On Sunday I kept busy reading some scientific journals and *The Sheltering Sky* by Paul Bowles. By Monday I was pretty sure I'd go insane.

Tuesday finally arrived, and I waited with my bags packed, giddy with excitement, a prisoner about to be sprung. Morning turned to afternoon and still Bill and Bob hadn't shown up. Finally the phone rang. "Where are you guys? When are you coming?" I asked.

"We're dealing with a million things. We'll be there as soon as we can, Don," Bill explained.

"Okay. Hurry."

I paced the room, stared out the window, tried to read, then paced some more. Around five I went down to the hotel bar and had a cup of tea while the sky turned dark. I heard the first rumbles of thunder, and a few minutes later a torrential downpour drenched the city. I finished my tea and went to the lobby to wait, growing more and more apprehensive. Driving in Africa at night can be dangerous enough; in a blinding rainstorm anything can happen. As lightning illuminated the landscape, I looked out the door and could see that Addis had already come to a standstill—the little Fiat taxis were flooded, and even the packed public buses were unable to move in the rapidly rising water.

Bill and Bob, soaked to the skin, finally burst into the lobby. "You're here!" I cried, leaping up. Eagerly I motioned for a porter to bring my luggage from my room, and we joined Yoel in our nice new vehicle with its heater blazing. Bill skillfully navigated the dangerous roads for the 100-kilometer drive, which took almost two hours. Starved for conversation, I enjoyed the ride, reveling in excited speculation about what fossils awaited us in the field.

"The Adama Ras is up here on the right somewhere, isn't it?" Bill asked when we entered Nazret, a sprawling town along the main highway. As he slowed down, we all leaned forward, squinting in the dark and the rain.

"There." I pointed, and he turned in the entrance, coasting down the long, gravel-lined, potholed driveway toward the dim light of the front office.

Inside a clerk wearing a warm-looking *gabi* greeted us with a bored expression. When I told him my name, he consulted a logbook, then reached into the drawer of the reception desk and pulled out a key. "Number four," he announced.

I nodded. "Fine."

We followed him up the stairs, and when he opened the door of the room, my heart sank. About twelve feet square, it had turquoise walls, mustard trim, and orange and white striped curtains. It was bone-chillingly cold and lit with ugly fluorescent lights. I accepted the key, thanked him, and glowered at my chuckling companions. Going over to the bathroom I flicked on the light and watched a dozen cockroaches scatter for cover under the toilet base. There was a shower but no shower curtain.

"Just like the Ritz," observed Yoel.

"I've slept in worse."

We went to the restaurant and had Cokes. I tried to extend the visit, even though I knew they had a long ride back. But pretty soon the rain let up, and they started looking weary. Reluctantly, I rose when they did and walked them out.

Bill shook my hand. "Make the best of it, Don. Before you know it, we'll be on our way to Hadar."

I watched the red taillights of the Land Cruiser disappear down the driveway, then returned to the restaurant for lasagna and a cold beer. With an unhappy sigh, I went back to my room, sat on the horribly too-soft bed, and knew my next chore was mosquito duty—they hide in curtains and in dark recesses like the plywood closet. From the bathroom I grabbed a towel and rustled the curtains, closet, and corners of the room, killing as many as I could. Luckily I was exhausted and was soon sound asleep.

At 1:30 a.m. the raucous barking of feral dogs roaming the streets in search of food—not an unusual occurrence in Africa— woke me. On this night the cacophony was disconcerting, because it sounded like the dogs were just outside my window. I turned on the light, swallowed half a sleeping pill with my saliva, lay awake for about an hour, then fell asleep again.

At seven I heard the sounds of the city beginning its day. My back was aching. I turned on the shower, grateful to discover the water was nice and hot. In my suitcase was a good bar of soap and a fluffy, absorbent towel. When I stepped into the shower, I looked down and saw that my legs were covered with fleabites.

Back in exile, I dreaded the next who-knew-how-many days at the Adama Ras. I was used to roughing it—sleeping on the ground in a fly-infested tent, then showering outside on a flimsy wooden platform while just inches below my feet a dozen creepy-crawly scorpions enjoyed the shower, too. And I never minded the hard work of toiling in the hot Hadar sun until flames of pain shot through my back, then hunching over paperwork, bleary-eyed, until dawn. But I was no match for day after day of profound inactivity. My one friend was Eshetu, a waiter who must have felt sorry for me because he kept bringing me fresh avocados from his garden. "Everything okay, Professor?" he asked about a hundred times a day.

There was a pool behind the hotel, but it was a breeding ground for mosquitoes, not a refreshing swimming spot for humans. I spent a couple of depressing afternoons watching yellow-fronted bee-eaters perform acrobatics above the water. Overhead, Russian MiGs headed north, where the Ethiopians were engaged in an all-out war with what is now Eritrea, as well as the TPLF, the Tigrayan People's Liberation Front.

The phone lines to Addis weren't always working, and since there were no phones in the rooms at the Adama Ras, each day I loitered forlornly at the reception area waiting for an available line. News from Addis was sporadic, and I pictured everyone being productive—chatting, laughing, working nonstop to make final preparations to leave. With no one to talk to and tired of eating pasta twice a day, I was ready to break out. What burned me most was the feeling that I'd been deliberately deposed.

I managed to get through to the Ethiopia Hotel on the night of October 5 and caught Bob just as he returned to his room after dinner.

"Hey, Don, how's it going?" He sounded tired.

"Don't ask! How are things there, are we making progress?" I heard him sigh.

"Well, I have some bad news. As you know, Bill's dad has cancer. And it's gotten bad, so he's in the hospital now. Not expected to make it."

"Oh no! Does he have to fly back to the States?"

"He's leaving tomorrow. Bad timing, huh?"

"Really bad! Well, tell him to do what he needs to do, and to take care of himself. We'll be waiting for him at Hadar when he can make it back. Tell him I'm sorry about his dad."

"I will."

"So . . . any idea when . . . ?"

"We're closing in on stuff and will probably be able to pick you up on Sunday."

"Thank goodness! What time?"

"Not sure. As early as we can. We're anxious to get this expedition going."

"Fantastic." There comes a point in the field-research process—after you've flown thousands of miles, coordinated a team of scientists, repaired vehicles, bought supplies, and worked your way through a seemingly endless list of things to do—when you just say "Let's go," even though a few details still need to be ironed out. You reach that moment when you can't stand one more night in a hotel discussing and preparing; you just want to get on the road. I was glad Bob had reached that point. He was a well-seasoned professional—I had worked with him since the 1970s—and I had no doubt he'd pull it off.

"See you on Sunday. Tell Bill I'm sorry," I said again.

"I will. See you."

I hated hanging up. When I turned around, I saw Eshetu's smiling face.

"Everything okay, Professor?"

One more pasta lunch, one more pasta dinner, and I awoke on Sunday with the usual flies crawling on my face. Like a kid on Christmas morning, I jumped out of bed.

After a shower, I went to the restaurant. One thing I've learned is that it's smart to eat a hearty meal before leaving for an excursion, because often you don't know when you'll see food again. That morning I had eggs, ham, toast, and coffee. Then I went back up to my room. A knock on my door interrupted my packing. When I opened it I saw Eshetu.

"Phone call for you, Professor."

"Thanks." I followed him to the lobby. "Hello?"

"Don, it's Bob. We're all set to go, just have to fuel up."

I consulted my watch. "So you'll be here around one?"

"Should be. See you then!"

"See you!" I hung up, excited that I was about to be rescued. Back in my room I finished packing. At 1:00 p.m. I went to the reception area and waited for half an hour. Then I returned to my room to make sure I had everything and halfheartedly reviewed some paperwork. At 2:00 p.m. I went back to the lobby. No Bob. There was no way could I spend another night at the Adama Ras. At 3:00 p.m. there was a knock on my door, a friendly reminder that I had to vacate the room. I carried all my bags to the lobby, and just as I got there Bob appeared.

He greeted me. "Well, here we are, a little late, but we made it out of Addis!" I laughed, so relieved I almost hugged him. With Bob were two Ministry of Culture and Sports Affairs representatives. All expeditions in Ethiopia required a representative from the government to oversee the research, to ensure that all antiquities regulations were being followed, to assist with any bureaucratic problems that might arise in the field, and ultimately to provide a report to the ministry on all details of the season. We counted on the rep to speak with some authority to police, soldiers, or any local officials we encountered and help smooth the way; he was also in charge of the ministry permit stating our intentions and listing all foreign members of the expedition, along with their nationalities and passport numbers. An expedition usually requires a single rep, but this year we were assigned two young men with a professional interest in archaeology and paleoanthropology: Tamrat Wodajo and Tekele Hagos.

"How's the food here?" asked Bob.

"Pasta's good," I replied with a rueful chuckle.

By the time we finished eating, our other vehicles had arrived, along with the rented lorry packed high with the bulk of our field equipment. We discussed the possibility of spending the night at the Adama Ras and getting an early start in the morning, but all the rooms were booked. So at five in the afternoon, we headed off and reached the town of Metahara two hours later. Our rooms at the roadside hotel there made the Adama Ras look like the Hilton. But after a decent spaghetti dinner, we crashed; I slept in my clothes, a *gabi* over my head to provide a layer between my face and the battalion of mosquitoes. Everyone awoke early to the sound of roosters crowing, and within half an hour our caravan was on its way again. By now the rain had stopped, and the sky began to brighten: the dawn of a long-anticipated expedition.

At Awash Station, a regular stop on the Djibouti–Addis Ababa railroad, we had a quick breakfast of bread, scrambled eggs with spicy green peppers called *karia,* and strong coffee at the Park Hotel, then hit the road again.

"Excited?" Bob grinned at me. I grinned back, nodding. There was no need to explain to him the surge of ebullience I felt as we headed toward Hadar. On the way to a field site you always have a sense of promise. I'd had the same sense on the morning I abandoned my paperwork and found Lucy.

Now, driving through the familiar and beautiful landscape of dark, cindery lava fields, bright green vegetation, and extinct volcanoes, I was glad to see it hadn't changed much since my last brief visit to Hadar in 1980. Neither had the ferocity of the 110 degrees Fahrenheit sun that forced everyone to hastily roll up their windows so Bob could flip on the AC full throttle.

"The only thing we might need to worry about is fuel shortages," said Yoel. He was a direct but gentle man who, perhaps because of his time in the Israeli army, conveyed a sense of competence and experience. The Afar tribesmen called Yoel Dr. Gahola (Dr. Fossils), because he spoke incessantly about fossils and he would give them medicine when they were sick.

"I've been hearing about it," I affirmed. "All over Ethiopia, but especially in the remote areas."

"Right."

We were all relieved when we pulled into the Total Station in Gewane and found plenty of diesel.

"Isn't there a funky little café around here, too?" I asked.

"As I recall, pretty good spaghetti!" Bob said. We laughed, but after our tank was filled, we drove to the café and that's exactly what we ordered.

As soon as we began to eat on the veranda we were joined by a gangly ostrich that had as much interest in our food as we did. With admirable determination, this fearless, comical bird made it his business to steal food off the plate of anyone who wasn't vigilant. His antics kept us amused throughout the meal.

After lunch, we stretched our legs, then got back into our Land Cruiser and headed to our next destination, the little town of Mille, some 160 kilometers to the north. The asphalt road, built in the early 1970s by Trapp, a German construction company, showed quite a bit of wear and tear, and Bob had to carefully maneuver our vehicle around deep potholes. Serving as the lifeline between the port of Assab on the Red Sea and the landlocked capital of Addis Ababa, the road took a lot of abuse. Huge Fiat double trailer trucks made the trip a tedious procession: Those headed south were heavily loaded and fumed big black clouds of diesel smoke, while the empty ones going north carried the rear trailer on the truck bed, which made the vehicle seem two stories tall.

After an exhausting three-hour drive, we rolled into Mille late in the afternoon. Not much had changed since we'd been there in the 1970s—it was still a ramshackle assortment of small structures hammered together with sheets of corrugated aluminum and plywood. Aimlessly roaming the unpaved main street were people dressed in all sorts of interesting outfits that combined American clothes and traditional garb. Occasionally the unmistakable horn of a large Fiat truck could be heard as it barreled through town, leaving a cloud of dust that totally obscured everything—goats, sheep, donkeys, and people—in its wake. Our conversation had dwindled, and everyone felt sluggish and sick of being in the car. My halfhearted offer to take over the driving had been rejected by Bob with a flip of his hand, thank goodness, so I leaned back in my seat with my eyes almost shut. Suddenly I sat bolt upright.

"Did you guys see that?"

"See what?" Bob asked, turning his head toward me.

"I'm not sure . . . we have to go back! Turn around!"

"Why? What was it?"

Stubbornly refusing to divulge what I thought I saw in case I was wrong, I insisted, "We have to turn back . . . it'll only take a second."

Bob pulled over, monitored oncoming traffic, then headed back.

"There," I said, pointing. On a flattened white metal sheet, someone had painted an unmistakable depiction of the Lucy skeleton. The metal sheet wasn't tall enough for the whole skeleton, so the leg was simply moved up next to the chest. She was such a popular icon throughout Ethiopia that Mille wanted to be identified with her discovery. Bob stopped the Land Cruiser, and with renewed energy we got out and went over to admire it.

"Who in the world would have done this?" I wondered, pleased and mystified. I pulled out my camera, and within just a few minutes we were surrounded by children who wanted their picture taken with Lucy.

Smiling and feeling welcome, we got back into the Land Cruiser. Our next stop was the Efrem Hotel for a cool drink. Once we'd had a chance to unwind a bit, Tekele said, "Don, as leader of the expedition, you have to check in with the local government officials and present your permits before going to Hadar."

"Of course. Where do we go?"

Tekele had already scoped out the lay of the land, so he led us on foot to a nondescript building that was the seat of the local administration. We were greeted by a man wearing a cloth skirt and a turban who told us to go see Hailum, the local party chief of the Mille region.

"Where is he?" I asked.

He shrugged and returned to his duties. We asked around and finally found Hailum drinking tea at Efrem's.

"Don Johanson," I said, then swiftly introduced him to the others in my group. He stood, shook my hand, and invited us to join him. Hailum promised to arrange everything by the next morning.

My night at the Efrem Hotel consisted of little sleep but lots of mosquito killing; all the while I was trying not to move too much in the stifling heat. In the morning an unusual sound woke me: baa-aaa. I sat up and looked into the eyes of an inquisitive goat standing in my doorless entryway. With a groan I rose and tried to work out the kink in my back; then I went to meet the others for breakfast.

"You must go to Dubti to meet with the first secretary of Assab," Hailum announced.

"How far is it?"

"About seventy kilometers to the north."

My willing nod belied the impatience I felt; it would probably mean another night at Efrem's, but I had no choice.

We went to Dubti, tracked down the first secretary in a crowded office, and retreated while Tamrat, whom we nicknamed "the ambassador," spoke with him sotto voce. After a short back and forth, the first secretary looked me directly in the eyes and said matter-of-factly, "Tomorrow morning at seven, in Mille."

"Okay." I shook his hand, but my hopes were fading. Dammit!

We climbed into our Land Cruiser and drove back to Efrem's. No one spoke; we were all tired of the delays, but there was nothing to be gained by complaining. At least this time you have company, I consoled myself.

As we pulled into the dirt lot in front of Efrem's, a small van pulled up next to us. It had a pastoral scene of the Awash River painted on its side, and the word LUSI.

"She's famous!" Bob laughed.

Out of the van stepped Efrem himself: short, stocky, and disheveled. I hadn't seen him in ten years, and although he looked much worse for the wear it was unmistakably him. We greeted each other with a warm embrace.

"Join me in a *bunna*," he said.

We were happy to participate in the elegant coffee ceremony: Women roasted and ground the beans by hand in front of us and served the coffee in little espressolike cups—with tons of sugar unless we stopped them. We passed a pleasurable hour catching up with him, and he told us over and over how excited he was to have "the Lucy team" staying at his hotel.

The mosquitoes and the goat and I made it through another night, and at precisely seven the next morning Tamrat, Bob, and I presented ourselves at the administrative office.

"So where is he?" Bob whispered.

"He'll be here," I whispered back.

We waited a few minutes, then heard a racket outside and went to see what it was. A caravan of Toyota pickups sporting 50-caliber machine guns in the back roared past, like a scene in a movie, with the first secretary in the lead vehicle. He spotted me and motioned for us to follow.

"Let's go!" shouted Bob, and we ran to our Land Cruiser. Bob gunned the engine and, enveloped in a cloud of dust, we took off in hot pursuit.

"What the hell!" I exclaimed.

"Don't worry, we're fine," Tamrat assured us.

"If you say so!" I said grimly.

At top speed we followed the first secretary to Odaitu, a government-supported rest stop and service area for truckers supplying Addis Ababa. Breathless from the ride, we entered a large building with a bustling cafeteria and were quickly ushered into an adjacent room. Inside more than a dozen officials, including a four-star general and the minister of transportation, were eating breakfast. I was overwhelmed and intimidated. We were told to sit; uneasily, we began to eat with them.

The mood was tense. For several minutes there was silence except for the sound of chewing. At last the general began to ask questions: What was the purpose of our research? How long would we be in the field? When did we intend to go into Hadar? How many *ferenges* were in the group? How do we interact with the Afar tribal people? And so on. My prompt, respectful answers eventually softened his demeanor, and everyone relaxed and enjoyed the meal of fried goat and french fries. Then suddenly it was over. As one group, the officials rose, said how good it was to meet the Lucy discoverer, and walked out.

After recovering from our surprise, we left, too, and got into our Land Cruiser. The party chief, Hailum, asked for a ride back to his office. "Sure thing," said Bob. We pulled out and headed back in the direction of Mille.

"Tomorrow morning," Hailum announced as we dropped him off, "be prepared to go to the field, or back to Addis."

Whoa! What? I held back a groan. Could we really have come this far only to have the rug pulled out from under us at the last minute—just when I could almost taste the fossil-rich Hadar deposits?

"Try not to worry," Tamrat said diplomatically over drinks at Efrem's. "For strategic reasons, the central government is concerned about this part of the Afar, Wollo province."

"Why? What's up?" Yoel asked.

"The TPLF is closing in from the north. They're hoping to liberate Ethiopia from the stranglehold of Mengistu, and in order to accomplish this, they have to fight their way through Wollo to reach Addis."

"Are they by any chance going to be using the road that cuts through the Afar?" I asked.

Tamrat nodded and explained that the government was going to protect that road at any cost. So if our team got the go-ahead to work, we'd be living in a war zone for two months. The news shocked us. All we wanted to do was get to the field and find some fossils.

"This is life in Ethiopia now." Tekele sighed.

At four thirty, one of the Toyota pickups with the machine guns swung into the hotel's lot. Out stepped the first secretary; he wore a blue safari suit and was smoking a Rothmans cigarette. We rose, invited him to join us, and ordered a coffee for him. "Well," he said, "we have a solution."

"What is it?" I asked, trying to stay calm.

He scratched his dark curly hair and took a deep drag on his cigarette. "You will be accompanied by a military platoon into the field."

Oh, this was bad. How could we show up in the village where all of our Afar friends lived with a platoon of soldiers?

"I don't think that's necessary," I said carefully. "The Afar protect us. They always have."

The first secretary's displeasure was evident; he shrugged dismissively. "If you want to go into the field, you must take soldiers."

"In that case, thank you, I appreciate your concern," I said, sounding more gracious than I felt. "When do we leave?"

"When the troops are ready," he answered. He rose abruptly, shook my hand, and went back to his vehicle, which roared out of the lot, headed back to Dubti.

"This is outrageous," Bob said. "The Afar are so independent. So proud. They distrust anyone who isn't Afar."

"I know," I said.

"They're going to hate this!"

"I know."

"Don, we can't just show up with—"

"I know, I know!" I finished my now-cold coffee in a single sip. "We have to go to Eloaha and let them know. We have a lot of friends there. I think they deserve to be told what's going on."

"Good idea," agreed Yoel. "That's the right thing to do."

"Make sure you get Hailum's permission," Tamrat said.

I was feeling overwhelmed.

Hailum gave us the green light to visit Eloaha, the last village en route to Hadar, the next day, but only on the condition that he accompany us. I had no objection; in fact, I thought his presence might be helpful. As I lay in bed that night I wondered what conditions I would find in the village after such a long hiatus; who was still alive? Our success depended a great deal upon the cooperation of my friends, two in particular. The first was Meles Kassa, whom for years we'd mistakenly called "Melissa." A charismatic and deeply intuitive Christian Tigrayan, he had, for some reason, chosen to settle in a Muslim Afar village, where he became so highly respected by the Afar that they made him an honorary member of their tribe. Linguistically blessed, he'd mastered the Afar language and spoke his native Tigrayan, as well as Amharic; occasionally we even conversed in Italian. In the past he'd been an invaluable member of my expeditions, serving as interpreter, political adviser, negotiator, and head of our excavation teams. Universally respected in Eloaha, he was frequently consulted to settle differences between Afar and non-Afar.

The second person was Mohamed Gofre, a revered Afar elder with a full appreciation of the significance of our work, which he had masterfully communicated to his fellow tribesmen. He and Meles had been hardworking, dependable participants in many of our expeditions during the 1970s. Their support, understanding, and powers of persuasion were essential to our success.

That I needed to meet with them before we arrived with a bunch of soldiers was my last exhausted thought on that sticky, humid night in Mille before I rolled onto my side and drifted off to sleep. The sound of mosquitoes buzzing and the muffled voices of other guests drifted in from the darkness outside.

Right after breakfast we set out, heading west toward Eloaha. I was excited, but at the same time apprehensive. Without telephone or postal service we'd had only sporadic and not always reliable reports filtered through the grapevine about happenings in Eloaha; we never knew for sure who had passed on and who was still alive.

Coasting slowly into town, we stopped at a little restaurant called Sentaiu's, where we noticed a group of curious Afar tribesmen observing our arrival. When I climbed out of the vehicle, I felt the back of my shirt soaked with sweat. A dozen or so children circled me, and one of the older kids said, "Johanson!" Everyone started to smile. Two little hands found their way into mine, and they all accompanied us into Sentaiu's for a cool drink.

News of our visit traveled quickly through the small village of several hundred people, and within just a few minutes my old friend Meles showed up. Looking older and thinner, he loped over and embraced me, then peppered me with questions about people he'd met when we worked together. He said how pleased he was that the Lucy team had returned to the Hadar site; he knew very well that our presence would bring work and salaries to many in the village, which was a good thing for everyone.

While sitting on the veranda of the small restaurant, I suddenly caught sight of a well-coiffed Mohamed Gofre slowly making his way toward us. Both alive! I thought with relief as I rose to greet him. Dressed in a new cloth skirt that he must have held in reserve for a special occasion such as this, he also wore a carefully folded white cloth over one shoulder. Striking the familiar pose of an Afar warrior, his head held high and proud, he carried a rifle. His waist was adorned with an ammunition belt housing an impressive row of bullets.

I solemnly held out my hand, which he accepted into his own slender hand and kissed the back. I returned the gesture. Then he squeezed my hand very slightly and placed it over his heart, and I did the same. This brief ceremony was the crystallizing moment of being back in Afar country. This was the intimate Afar greeting, one of trust and respect.

We were joined by more and more familiar faces and caught up on the news of those who had passed away during our absence; some had died from undiagnosed and untreated diseases, while others had been killed in skirmishes with the Issa, a rival tribe that was encroaching from the south. The Afar don't show much emotion about the departed: Once someone is dead and gone, he or she is, well, gone. As devout Muslims, they believe that the departed are in the good hands of Allah. In fact, when I gave Gofre a photo of himself standing next to our now-deceased friend Abraham, who had worked with us in the 1970s, he carefully tore off Abraham's image, handed it to me, and accepted what was left of the photo.

After a brief visit, Meles and Gofre decided to accompany us back to Mille, which gave me an opportunity to explain the delicate situation with the soldiers.

"We have no choice," I said. "If we did, believe me, we wouldn't bring them."

"We understand," Gofre said with a glance at Meles, who nodded. "The government has a job to do."

"It won't be a problem," Meles concurred, "but we would urge you to make sure these soldiers don't infringe on the privacy or rights of the Afar."

"Absolutely, you have my promise," I said, cautiously optimistic.

After a final night in Mille we awoke early on Friday, October 12, to make the vehicles ready for the trip to Eloaha, and then to Hadar. We were set to go by seven, and Hailum escorted us to the soldiers' barracks, where he introduced us to Lieutenant Woube, a brash, baby-faced twenty-year-old in charge of twenty-eight men. Could this get any worse? I wondered as I greeted him respectfully, then watched as he hustled his men into a large Fiat truck.

After an arduous five-hour drive, we reached the banks of the Awash River around three thirty in the afternoon, at the height of the heat. I didn't have to tell my team that we had less than three hours to put up tents, get the kitchen supplies unpacked, and assemble some temporary tables for dinner before the sun went down. Arrival is always a scramble, but no one wants to sleep outside with the dense, possibly malaria-bearing mosquitoes.

Woube quickly mobilized his men, and they began to dig fox-holes around the perimeter of our encampment, after which they set up large machine guns. By the time we'd eaten and I'd crawled into my little tent, it was 9:00 p.m. I had major reservations about the entire setup. I had come all this way to find fossils, not to live in an armed camp with an obvious tension between my lowland Muslim friends and the officious highland Christian army militia! But as I dozed off I had the happy feeling that we would all be back in the field working. In spite of everything I couldn't lose sight of the fact that we had made it back to the very place where Lucy was found. I fell asleep feeling vindicated: After all the years of absence and trials and tribulations to get here I was finally back in Hadar.

CHAPTER 4

Pay Dirt

As I watched the sun rise over the lush forest on the far side of the coffee brown Awash River the next morning, I conceded that I might have aged somewhat since my last cup of *bunna* here, but Hadar seemed timeless. Except for the perpetual erosion of the rapidly flowing Awash, which forced us to set up our cliff-top camp a little bit downstream from our usual spot, everything else looked pretty much the same. Behind me the landscape was barren, a desertic area exposing layers of rock, a stark contrast to the verdant riverine forest. I loved watching the vervet monkeys come down to the river for a drink, on alert for the resident crocodiles. The grunts of warthogs and the twittering of hornbills, Abyssinian rollers, and sunbirds filled the air. It sure felt good to be back in the field.

As if reading my mind, Bob leaned back in his camp chair contentedly. "Like we never left."

In comfortable silence, we listened to the sounds of a camp coming alive, and pretty soon Wubishet, the cook's assistant, made the welcome announcement that breakfast was ready.

"Thank goodness, I'm starved," I said, then chuckled to myself. I certainly wasn't one of those mad-scientist types who has to be reminded to eat. Bob and I grabbed our chairs and joined the others at a makeshift plywood table balanced on cardboard boxes. I had enjoyed many wonderful breakfasts at Hadar, but that morning Alayu's fried eggs were as delicious as a fancy full-blown banquet.

On the first day all we wanted to do was to eat, get into the field, and find some hominids. But unfortunately, the demands of setting up camp took precedence, and so during breakfast we talked about what needed to be done and by whom in order to get the expedition up and running.

Bob volunteered to set up our water-processing system, because to drink from the Awash safely, we had to filter and chemically treat the water. Mike Tesfaye, an Ethiopian geologist who had for some inexplicable reason picked up the nickname Mike Marino, volunteered to set up the tent fly and the worktables. That meant Yoel and I would be digging a hole for the latrine and setting up the canvas-walled shower stalls.

We finished eating, then scattered to carry out our duties. As the sun rose higher, the thermometer dangling from my field bag registered 108 degrees Fahrenheit. I had forgotten just how brutal the heat was; the sweat was streaming down everyone's faces. At eleven thirty we collapsed under the tent fly, hungry, weary, all attention focused on Alayu's movements in the kitchen. Arguably the hardest-working person in camp, he did not disappoint, and at noon he summoned us for a salad of cabbage, tomatoes, and onions, topped with a puree of avocado and garlic, along with bread and plenty of precious water. Every two weeks we would have to send someone to market in Bati in the highlands, two and a half hours away, to shop for food and other supplies.

During the morning other changes had taken place, too. Under the command of Lieutenant Woube, the soldiers had put up their pup tents, and in the foxholes set on the periphery of the camp gun barrels gleamed in the sun, the metal probably hot enough to burn skin. Unfazed by the blazing sun, Woube strutted around in full uniform, clearly prepared for any surprise attack. Watching him rue-fully, I thought his presence should have inspired confidence, but the truth was, if the camp was attacked the only options were to leap over the cliff into the Awash River or get caught in the gunfire. It was as safe as any scene in an Indiana Jones movie.

After lunch I sat under the tent fly and noted in my journal that things seemed to be falling into place. It felt so deliciously familiar: the smell of the campfire, the sounds of camp, the anticipation of finding more fossils.

We spent the rest of the afternoon organizing our equipment and personal gear. The kitchen crew had thoughtfully put out the plastic sun shower bags to warm up and around six, as the sun began to set, I grabbed one and headed off to the showers. It was glorious to be free of two days of sweat and dirt, and as I dried myself in the twilight I looked up and identified Saturn, a bright dot in the southwest portion of the sky. I walked back to my tent in my towel and flip-flops, put on a clean pair of shorts, a long-sleeved shirt, and sneakers, and dabbed some DEET on my legs and arms before heading out to the table for dinner. "Sure is a lot of *bimbi* pressure," someone observed, referring to the flotillas of mosquitoes that inevitably closed in on us at dusk. I was glad I had remembered to take my antimalarial medicine.

After a supper of fried potatoes, cabbage, and canned corned beef, I formally welcomed the fifteen participants and distributed IHO lapel pins and T-shirts. "I would like to welcome you, too," announced Mohamed Gofre. "And I ask Allah to bless this expedition." With the others I respectfully bowed my head. I was a lifelong atheist but certainly not inclined to turn down a blessing! It wasn't long before the team settled into their tents, and the last sounds I heard before falling asleep were the splashes and grunts of hippos in the river.

After breakfast the next morning, I presided over a half-hour briefing. We were going to be at Hadar from October 12 to December 15, two full months of fieldwork. But I reminded the team that time would pass a lot faster than expected, so we would have to maximize our efforts every day. It was an easy sell and there were nods all around—they were here to work. Breakfast would be served at six thirty, which meant getting up at six and being ready to leave camp at seven thirty. We would work until noon, the hottest time of the day, then head back to camp for lunch at one. Team members who were working farther out in the field would bring lunch with them and eat it there. "A morning in the field is hot, hard work," I reminded everyone sternly. "Don't try to be a hero. Drink lots of water, and after lunch, rest for a couple of hours. If anyone wants to go back to the field in the afternoon, I strongly recommend that you wait until four." Heads bobbed, and I knew everyone wanted to get going. I was anxious to leave, too, but I felt it was important to provide a perspective on what had already been accomplished at Hadar, why we had returned, and what the major goals of the current field season were. After all, some participants, like Yoel Rak and the two government representatives Tamrat and Tekele, had never been to Hadar. The main objective of my briefing was not to go into excruciating details of our strategic scientific plan, but to lay the ground rules for fossil collection, geological work, and the daily schedule of camp life.

I explained that our focus in the field was twofold. As a team, we would conduct a systematic search for fossils, especially hominid, although certainly we hoped to collect rare nonhominid fossils, such as carnivores and monkeys, as well. Then I pointed at the first-timers: "Each of you will be expected to recover at least one hominid skull." There was startled silence, followed by laughter. I grinned, then got serious again, noting that we would record fossil concentrations and assess erosion at localities we had excavated in the 1970s. We would also make a sweep of previously collected hominid localities on the off chance that further fragments had surfaced.

It was vital that no one collect a single fossil before its exact position was recorded and notes were made on the circumstances of discovery, I continued. A fossil taken out of context had very little value to us, and, if improperly collected, it might even be misleading. We had to have the geological context and geographical position in order to correlate the find with other discoveries. The geology team would work under the direction of Bob Walter, whose objective was to resolve the age of the Hadar Formation, especially the deposits where Lucy had been found.

"All right," I said. "Let's collect our gear and get out of here!"

Our first stop was a small area within walking distance called Hominid Valley. It was here in 1974 that Alemayehu Asfaw had distinguished himself by finding some hominid jaws, and I invited him to share with the team what that experience was like. Then I asked Bob to talk about the local geology.

We walked a bit. Pausing at the top of a little knoll, I asked Tamrat, "What do you see? Describe it to me."

"Lots of dirt," he replied, somewhat unenthusiastically. Suddenly he blurted out, "A tooth!" Then he knelt down and pointed at a two-inch-long rectangular specimen. I crouched next to him for a quick inspection. "It's the upper molar of a hippo," I observed as I lifted it out carefully and handed it to him. Referring to the topographic map on my clipboard, I announced, "This is A.L. 200. If we were here to collect, we'd put this into a cloth bag and label the attached yellow flap A.L. 200, October 14, 1990, and initial it TW, Tamrat Wodajo. But for now, let's leave it here."

"It's interesting to note," Bob added as Tamrat replaced the tooth, "that we're in the Sidi Hakoma Member in one of the lower sand horizons, roughly equivalent to some of the oldest layers at Hadar."

"How old?" asked Tamrat.

"Approximately three and a half million years."

I sensed everyone's reaction of amazement and could feel them retreat to their private fantasies. We all wanted the same thing: The Big Discovery. You can't be in the field and not experience that pull. We scoped the area, and I helped identify a variety of bones that Tamrat, Tekele, and the Afar found: Hominid? No. What about this—hominid? Nope. What about this? No, sorry. After everyone's hopes had been dashed, I whistled for the wandering Alemayehu to rejoin us, and we trudged back to camp for a drink.

"How about a trip to the Sidi Hakoma Tuff?" Bob suggested, referring to the bottommost layer of volcanic ash in the Hadar Formation.

I nodded. "Good idea." The area in and around camp is referred to as the Sidi Hakoma, which means "three hills" in Afar, and the tuff is a light buff-colored layer of volcanic ash, which is best viewed near another hill known as Kenia Koma ("mosquito hill"). Easy to spot throughout Hadar in the lower geological levels and just a short ride by Land Cruiser, it became one of our most obvious marker horizons. When we got there, Bob knocked off a chunk of rock with his hammer and pointed out the microlayers. He explained that what we were looking at was ash spewed out of a volcano millions of years ago that had been altered to a clay, and therefore could not be used for argon dating because the argon had been released into the atmosphere. If we could nail down a geological age for this ash, we would be able to definitively establish the antiquity of the lower portion of the Hadar Formation. He faced the group. "Does anyone know why the ash is so thick here?" No one answered.

Bob escorted us to an area nearby where erosion had exposed an especially thick section of the Sidi Hakoma Tuff. We were looking at an ancient river channel where the ash was so thick it had actually choked the flow of the water. The cross-bedding of the strata indicated turbulent river flow, he observed. Next he removed a chunk of the ash, which he handed to Mike, along with his hand lens.

"What do you see?"

"The microlayers are crosshatched, like you said."

"Right."

Turning and looking over his shoulder, Bob gestured with his rock hammer toward horizontal strata of darker sediments. "Those layers, clays and silts, were deposited in calmer water, most probably

a lake." It was a neat geological lesson that everyone grasped immediately: The depositional environment left a diagnostic signature in
the arrangement of the sediments. Those deposited in calm lake
water gave the appearance of a layer cake, while sediments laid
down in higher-energy environments like a river showed what is
technically dubbed "cross-bedding."

We climbed back into the Land Cruisers, and I headed north to
A.L. 333, where the First Family was found. What a thrill to turn off
the main drainage we were following and see the familiar landscape
in the distance! My thoughts flashed back to the mid-1970s, when we
had excavated the tons of rock and sediment that surrendered the
remarkable collection of *Australopithecus afarensis* from a geological
instant in time. I had always felt further work at 333 could yield additional hominid fossils and hoped during this field season to develop
an appropriate excavation strategy; actual excavation was not on our
agenda, however—the team was too small and there was too much
other work to be done. But if I couldn't work the site, the least I could
do was show it off. I knew that everyone, especially Yoel, was eager to
see where paleoanthropological history had been made.

Erosion had pretty much obliterated our 1975–77 excavations,
but I was glad to detect a faint outline of the work we'd done then.
"Did I mention it's a heck of a climb up to the site?" I said, indicating the hill. "But trust me, it's worth the walk." There was no need
to coax them; they were already starting up. I had to smile as I
watched their heads swivel, their eyes on the ground, all hoping to
spot an ancestral bone. And, of course, my own gaze was drawn to
the ground, too.

When we reached the summit, I caught my breath and thought
back to the thousands of man-hours that had been put into the excavation and screening that produced the First Family. Everyone stood
in awe, taking in the spectacular view, and I asked Bob to talk about
the geology of the hill. He obliged, explaining that we had left the
Sidi Hakoma Member and climbed into a younger level of the

Hadar Formation called the Denen Dora Member, named for the wild asses that once populated the region.

"Don, let me have your field glasses."

I handed them over. Peering through them, he focused on a basalt plateau in the distance known as the Kada Damum, "big face" in Afar. Others who had brought binoculars followed suit.

"That's one of our main geological problems," he said. "We don't know exactly where the Kada Damum fits into the geological sequence, and even if we did, we don't have an age for it—frustrating."

Heads nodded. In a lecture hall they might have had more questions, but right now they were just too eager to see the next stop on the Hadar tour—the exact spot where I'd found Lucy. Leading them back down the slope, I then drove eastward to intersect the Kada Hadar, the dry wash that gave the region the name Hadar. The precise meaning of Kada Hadar is unclear, but in Afar it is best translated as "big white water," perhaps a reference to when it floods during the rainy season. I didn't exactly remember where to turn off—there aren't any convenient landmarks like a Shell station or a Walgreen's—but the veteran Mohamed Gofre knew exactly where to leave the Kada Hadar and head north.

Soon I recognized the area, and I felt my heart pound faster: I pulled over and cut the engine. The energy of the place was exhilarating, and everyone quickly jumped out of the vehicles. Slowly I got out and joined the others: I didn't expect to feel such a powerful wave of nostalgia. I cleared my throat. There was no way to describe my emotions: the sheer ecstasy of being back at the locality that had defined my career. Then the intensity of everyone's anticipation snapped me out of my reverie, and with renewed enthusiasm I led the scramble up a hill of loosely packed lake deposits to a vantage point where we could look out onto the slope that had given us Lucy. In that moment I knew that every single team member was hoping to discover another skeleton. But all we saw were a couple of eroded piles of sediments, the remains of screening debris, and, as we walked closer, a rusted nail and a flaky piece of orange survey tape.

With a sense of wonder that had not diminished in almost sixteen years, I explained exactly what had transpired on that unforgettable morning in 1974. First I pointed to the precise spot where I had spied the fragment of ulna (elbow bone), then pointed upslope to where the other bones had lain.

Yoel stood with a look of disbelief on his face: "This is it? Right here? This place looks no different from any other spot we saw today."

I laughed. To me it looked completely different from any other place on Earth.

Bob took the opportunity to talk about the importance of accurately recording the exact location of a fossil. He led us over to a series of sediments that had been disrupted by a fault—the Lucy Fault, he called it. Which side of the fault the deposits lay on determined their age. If a fossil that had been picked up could not be precisely placed relative to the fault, it was possible to drastically underestimate or overestimate the specimen's age by perhaps hundreds of thousands of years, he explained. The group nodded, and those with cameras took pictures. Everyone was still hungrily scanning the ground for hominid bones. The urge was irresistible.

On the drive back to camp Tamrat and Tekele asked questions nonstop: So Lucy was just lying on the ground? No one else had seen her? And you knew right away she was a hominid?

"It's really overwhelming," Tamrat admitted.

"Don't worry," I replied. "Over the next two months you'll come to know Hadar as well as the rest of us."

A couple of days later, just after lunch, when camp was pretty quiet except for some activity coming from the kitchen, I sauntered over to the work tent, where Bob was peering through a field microscope at some samples he had collected during the morning. He heard me coming, looked up, and said, "Incredible, really incredible! Look!"

I sat and brought the binocular microscope into focus. "What am I looking at?"

"See those sharp-edged grains in the middle?"

"Clear as a bell. Are they feldspars?"

"Not just any feldspars! Sanidine feldspars! The best volcanic material for argon dating!"

My jaw dropped as the implications became clear: We might be able to get a precise age for one of the volcanic layers in the Hadar Formation, but which one? The feldspars, Bob explained, were found in a volcanic ash called the Triple Tuff located at the interface between the Sidi Hakoma and Denen Dora Members. Furthermore the Triple Tuff lies some 20 meters below the geological level that yielded the First Family.

"You're saying we're finally going to be able to figure out the geological age of the First Family?"

He nodded. "Getting closer." He politely maneuvered me away from the microscope so he could get another look. How gratifying, I thought, a breakthrough so soon in the season.

The date would turn out to be 3.24 million years ago.

Bill Kimbel arrived at Camp Hadar in the middle of November. I greeted him with a sympathetic hug and said how sorry I was to hear about his dad.

"Thanks."

"How are things at home—your mom okay?"

"Okay, under the circumstances. How are things here?"

"Running without a hitch."

"Good."

His gaze traveled past me and took in the activity: from the physical setup of the camp and Lieutenant Woube's incommodious presence to the general busyness of team members performing their duties, most of which centered on the pile of specimens laid out on the table in the tent fly. "Well, so, what have you found?" he asked.

"Come see." I showed him the few fragmentary hominid fossils we had collected, bits of jaws and teeth.

"No skulls?"

"Oh, right, I guess I forgot to show you the skulls," I teased, but I knew he was disappointed, and so was I. We all were. An *A. afarensis* skull would resolve the ongoing criticism from our colleagues about the anatomical distinctiveness of Lucy's species. We had fragments of skulls, but nothing close to complete.

He grinned. "Well, the season's only half over."

"Right!" I said. I could see that being here was just what he needed to move past his grief. And I was sure glad to have him.

Accompanying Bill were two other geologists who had come to assist Bob in sorting out thorny geological issues. Jim Aronson, Bob's Ph.D. adviser at Case Western Reserve University, had worked at Hadar in the 1970s and specialized in argon dating of volcanic rocks. He always brought a sense of levity and great optimism to camp. With his black-rimmed glasses and frizzy red hair and beard, he looked like a cross between a rabbi and Woody Allen. Paul Renne, a specialist in geological dating from IHO in Berkeley, had come to work specifically on issues related to the paleomagnetic record in the Hadar Formation. This was Paul's first trip to Hadar, and he was eager to get to work. Clearly a product of 1960s Berkeley, he dressed in a colorful tie-dyed T-shirt, which meant he really stood out since the rest of us wore drab khaki.

Although hominid finds were trickling in and the geological program was making great headway, there was an undercurrent of tension to the 1990 field season. Being surrounded by a contingent of soldiers watching our every move contributed to the edginess we all felt. But the most troubling concern was Lieutenant Woube's disregard for the Afar people. Out of respect for the expedition they tolerated him, but this was their homeland and we were their guests.

One day Dato Ahmedu, an Afar elder, complained to me that Woube had purchased goats from the rival Issa tribe across the Awash River. Aggravated, I explained to Woube that his behavior with the Issa was unacceptable.

"The last thing you want to do is get them mad," I said.

He held up his hand. "Okay, okay."

But just a few days later I returned to camp and saw Woube on the other side of the Awash buying another goat from the Issa tribesmen. Dato was furious and suggested that he and Tamrat go to the main village of Eloaha to consult with the tribal headman—the *balibat*—for spiritual and practical advice. I agreed and made preparations to have a car pick them up. Just as they were about to pull out, Woube and his Kalashnikov-brandishing soldiers rushed us. The lieutenant flailed his arms and shouted orders in Amharic and English in a high-pitched voice. The Ethiopian kitchen staff ran out to confront the soldiers, and the Afar assumed threatening poses—they, too, had Kalashnikovs. Suddenly a screaming argument erupted.

"Get out of the car!" Woube commanded Dato.

Glowering, Dato refused.

"Get out now!" Woube summoned his soldiers to surround the car. "If you try to leave, my men will shoot you!"

It felt as if I was in a nightmare: I was paralyzed by panic and unable to think of a single thing to say that would stop the insane events unfolding before me. Dato didn't want to appear weak and give in to Woube, but he had several assault rifles pointed at him. Fuming, he got out of the car. But he maintained his dignity, and I could tell by his gestures that he was describing to Woube in gruesome detail what the Afar tribesmen would do to the lieutenant as soon as they found out what was going on. Woube listened with increasing apprehension, then motioned for the soldiers to lower their weapons.

"Okay, I'm calling off my men," he said.

Dato glared imperiously at him, as if to say it was too late, the damage had been done.

"I called off my men!" Woube said, pointing urgently at the retreating soldiers who looked as scared as he did. "No more of this, I promise."

Dato pointedly evaluated all the armed men by balefully eyeing them one at a time. Woube added, "You're safe here. I won't let anything happen to you."

Everyone waited for Dato's response. He leaned on his walking stick and after several seconds delivered his final devastating line, which I asked Tamrat to translate for me: "My life is now in your hands; it is your responsibility." Woube nodded earnestly and escorted his soldiers away. I stumbled over to the dining tent and drew myself a glass of water. A stiff scotch would have been better.

Dato joined me, and with Tamrat translating, asked what I thought. I took his hand and said, "I have lost all respect for the lieutenant, and am very upset about the way he's treating my Afar friends."

Dato looked directly into my eyes and said, *"Yehe, mehe."* Yes, good. Reassured by my response, he turned and walked away. I was glad that disaster had been averted, but I wished cool heads and logic had prevailed, instead of death threats.

Our field season would end on December 15, and we still had much to accomplish before we broke camp. We couldn't afford to have our time at Hadar curtailed by any event. Of course, the dream of any expedition like ours is to find the specimen that resolves all the unanswered questions of human origins, but we stayed focused on the equally important issue of establishing unmistakably the geological dating of the Hadar Formation. The geological team hunkered down for an intense strategy meeting, during which they identified the remaining crucial issues.

"We won't leave until we figure out the stratigraphic position of the Kada Damum basalt in the Hadar Formation," Bob reported to me, "which means undertaking some detailed mapping in the eastern Hadar region."

I nodded—I knew what was coming.

"The problem is, our Afar men don't want to go with us because of all the Issa tribesmen in that area," Bob explained. "But maybe they'll go as long as they're told it's okay to turn back at the first sign of trouble."

"Good idea," I said.

The next day the geologists set out for an arduous climb up and down the steep, very hot exposures of black basalt. When they returned to camp they were elated.

"The basalt is in a graben," Bob declared. A graben is a large block of sediment, banded on both sides by geological faults, that has been displaced downward, forming a trench. Apparently in this area of Hadar the Sidi Hakoma Tuff sits some 30 meters below the basalt. Now details of the sequence of geological layers within the Hadar Formation were resolving, which would become more important when the basalt and other volcanic layers could be geologically dated.

"Fantastic!" I said. "One down, one to go!"

The next issue to be addressed was dating the Sidi Hakoma Tuff. Using only the traditional potassium-argon dating technique, we weren't able to date the tuff due to the scarcity of datable components. But now that we had the single-crystal laser-fusion technology, we just needed to find some feldspar crystals, the minerals appropriate for argon dating. Bob was certain that if he could crush and screen a large amount of the tuff, he could recover a sufficient number of feldspar grains for analysis. It was quite a scene: Bob and his assistants were perched on top of the Sidi Hakoma Tuff, enveloped in a swirling cloud of buff-colored dust as they crushed and screened some five hundred pounds of the rock. His energy and enthusiasm never faltered, and when he returned to camp he assured me that when he processed the residue back at IHO's dating lab, he'd have a sufficient number of feldspar grains for dating. This would give us the absolute oldest age of the Hadar hominids.

Bob knew that the tuff from every volcanic eruption has its own unique chemical fingerprint. In his 1980 Ph.D. thesis he noted that the Sidi Hakoma Tuff had the same fingerprint as a tuff from Lake Turkana that was believed to be more than 3.3 million years old. This meant that the Sidi Hakoma Tuff had to be at least that old. Back at Berkeley Bob focused his attention on the Hadar tuff and was able to date 20 feldspar crystals. Using the single-crystal laser-fusion method, he determined the age of the Sidi Hakoma Tuff to be 3.42 million years old. This was more or less what we all expected, but we were relieved that we now had a precise date. Two down!

In the middle of the afternoon on December 8, I was studying some of our hominid finds—fragmentary bits of jaws and teeth—and heard the sound of approaching cars. Since Hadar is not exactly situated near a major highway, it was always exciting to welcome visitors. Out stepped Berhane Asfaw, the director of the National Museum of Ethiopia; his supervisor, Tadessa Terfa, from the Ministry of Culture and Sports Affairs; Solomon Yirga, also from the ministry; and my colleague Tim White from Berkeley. I greeted them warmly. They wanted to find out how our research was progressing and get a personal tour of Hadar.

Over the next few days they toured Hadar, including the famous sites where Lucy and the First Family had been discovered, and we discussed the geology. I could tell they were impressed with the expedition's efficiency, but I also sensed some envy: Camp Hadar was a lot more comfortable than the digs in the Middle Awash, where Tim was working, and he had a long list of complaints: The food was lousy, there was no easy access to water, and the local tribesmen were not always friendly. In contrast, at Camp Hadar everyone on the expedition pitched in to make living conditions as pleasant as possible. We had a terrific cook, a convenient source of water, good showers, comfy personal tents, and we had enjoyed years of friendship with one another and the resident Afar. Visitors always raved about the accommodations, and we eventually earned the nickname "Hadar Hilton."

The tour concluded late one afternoon when Tim and I were

walking side by side and simultaneously noticed a bone eroding from a soft layer of sediment.

"Hominid humerus," Tim said.

He was right. Labeled A.L. 137-50, it was the lower three-quarters of a very heavily muscled right upper arm bone, the bottom end of which had been chewed off, probably by a hyena. The humerus, most likely from a male *A. afarensis,* was a welcome addition to the 1990 hominid collection, and promised to shed light on the suggestion that *afarensis* may have climbed in trees.

At breakfast on December 10, Berhane turned the conversation to the archaeological sites in the western region of our permit area known as the Kada Gona, named after a large riverbed that floods annually. In late 1976 Hélène Roche, a French archaeology student, had reported finding rudimentary stone tools made in the Oldowan tradition, so named for the primitive flake implements first discovered in Olduvai Gorge. She had found the tools on the surface, so she didn't know how old they were. And she had to return to Paris before she could put in a test excavation to determine the exact geological level from which they came. The geologists who had looked at the sites afterward had a hunch that the tools came from a level below the volcanic ash layer dated to 2.3 million years ago. But in order to know for sure, it was necessary to find some tools still in the ground. So I had invited my good friend Jack Harris, a stocky New Zealander with a passion for playing rugby, to come work at Gona.

We decided to dig in a promising spot that was at the same geological level but in a different geographical location from where Hélène had made the first discovery. In typical archaeological fashion we laid out a one-meter-square grid with string and began excavation. It didn't take long before we unveiled several square meters of the artifact-bearing horizon. These stone tools were in situ and they could not have eroded down from higher, younger strata.

"This layer might be two and a half million years old," I said, "and if that's the case, it will confirm that the stone tools are seven hundred thousand years more ancient than those from Olduvai Gorge in Tanzania." I was totally convinced that the Gona tools were the oldest yet documented. But I knew that Berhane and Tim wanted to see with their own eyes the stratigraphic position of the tools, and by inference estimate their geological antiquity. So, I took them there. Amid heated discussion about sloppy geology and incompetent excavation, Tim and Berhane continued to challenge the age of the tools. "Whoa," Tim said suddenly. Crouching, he pointed at a scatter of artifacts in close association with a volcanic ash horizon. His discovery left little doubt about the stratigraphic level of the stone tools. Examination of the flakes revealed the striking platform—a flat area where flakes are struck off a core with hammer stone.

I could see the rippled area immediately below the platform—the bulb of percussion—which is characteristic of purposeful stone tool manufacture. And the flake edges were fresh and sharp, suggesting that the artifacts had not been transported for any distance. They were real, their geological context was certain, and if argon dating provided an ancient date, these might just be the oldest stone tools ever found.

"So who should undertake excavation of the Gona site?" Tim asked that evening after dinner.

"Archaeology isn't a major research focus for our group, so we're not the appropriate choice," I said.

Jack had his sights on Gona, but I was certain Berhane would never let him have it—he had taken a strong dislike to Jack when they met back in the eighties. Sileshi Semaw, an Ethiopian national and a student of Jack's, was Berhane's choice. IHO had provided funds for Sileshi's graduate work, and he appeared to be a promising young scholar, so I was in favor of this arrangement.

"He's welcome to conduct his research from Camp Hadar," I said. What a great solution! Training Ethiopian scholars and involving them in primary fieldwork was one of my goals that was now mandated in the new regulations.

Perfect, Berhane declared.

Finally, we all agree, I thought.

The next day Tim and Berhane left Hadar, and camp life resumed its frantic pace. With only four days remaining, the pressure to complete our research agendas was relentless. Bill and I worked at breakneck speed to organize the 12 hominid fossils and 250 fossils of other vertebrate we'd collected, photograph them, enter the information into our computer database, and prepare them for transport to the National Museum in Addis Ababa. Bob and Jim's geology team was in a dither—Bob wanted to collect more volcanic rocks (he never seemed to get his fill), and Paul needed to complete his collection of sediment samples for paleomagnetic dating. With so few days left, there was no time for anything but work. Even meals were brief: rows of distracted bodies with mouths chewing and brains wondering how they were going to get everything done.

Paul had just recently become a research associate at IHO, and he earned an invitation to join the expedition because of his expertise in paleomagnetic dating. He explained the concept to some of the IHO students: The earth's magnetic field has not always been normally oriented as it is today; it flip-flopped many times when north and south exchanged places. Fortunately, when layers of sediment are deposited, the iron-rich minerals are oriented along the magnetic field and record the polarity at the time of deposition. The process by which the magnetic field changes isn't fully understood, but it's related to convection currents in the earth's liquid iron-nickel outer core. The earth's geomagnetic polarity time scale is calibrated with argon dating, and the part of that time scale we understand best is the last 10 million years.

Paul hoped that the combination of paleomagnetic data and argon dates would clarify the chronology of the Hadar Formation. He and Mike Marino had meticulously carved little pillars of sediment every three-quarters of a meter throughout the geological sequence. They had cemented a one-inch-diameter quartz-glass cylinder over each pillar and scribed magnetic north on the glass. Back in the IHO lab using specialized equipment, Paul would be able to ascertain whether the sand, clay, silt, or even basalt had been laid down during a "normal" or "reversed" period of magnetism, thus reconstructing the geomagnetic history of the Hadar sequence, which could then be correlated with the earth's geomagnetic polarity time scale.

Paul's remaining challenge was to sample the "rock hard" Kada Damum basalt, and for this task he had brought along a specialized diamond drill bit that was attached to a gasoline-powered drill. Even with this powerful tool he expended lots of sweat and effort to penetrate the basalt and recover a series of core samples for paleomagnetic evaluation.

We spent the final days in camp packing and taking inventory of all equipment and figuring out what needed to be repaired, replaced, or augmented. We chose Thursday, December 13, as the night to have our formal farewell dinner. Alayu prepared spaghetti with meat sauce, cabbage salad, and fruit cocktail for dessert. And with beer that had been cooled in the river, we toasted to a very successful season. We hadn't found an *A. afarensis* skull, but we had added to our hominid collection and, most important, collected critical data for the dating puzzle. After dinner everyone crowded under the tent fly for the final remarks.

"Huge thanks to our kitchen support staff," I began. Grateful applause broke out, and the kitchen staff was relieved; they were looking forward to returning to their families and friends in Addis Ababa for a much-needed rest.

"And, of course, it's always hard to put into words the appreciation we feel for our Afar friends. Thank you for having us as your guests." More clapping, which the Afar acknowledged with humble nods. "Although we live far away, we'll bring home special memories of our time together, and look forward to coming back next year." Mohamed Gofre gave a long speech that was translated by Meles

Kassa, who had worked with me since 1973 and who spoke fluent Afar. Meles's delivery was very dramatic, complete with sweeping gestures and colorful hyperbole. With tears in his eyes, Mohamed thanked us for honoring his people by coming to their homeland and providing a source of income that allowed them to obtain medical attention and buy clothing and food.

A few of us moved our camp chairs out into the dark and recapped the field season while we watched the Geminid meteor shower overhead. We could laugh now about the setbacks that had thrown everyone into a panic, the occasional squabbles, being hot, being hungry, being thirsty, being sore, being exhausted. We talked about how good it would be to sleep in a real bed, take a bath, eat in a restaurant. But as I retired to my tent, I started to feel the usual end-of-season despondency. As hard as camp life was, nothing in the world compared to the challenge, the fulfillment, the anticipation of a major find. I dropped off to sleep, only to waken at two thirty to the chilling howls of hyenas.

Everyone was up before dawn on December 15, exhausted after taking down camp the day before, and by eight thirty all the cruisers were packed to capacity and the roof racks piled high. With the lorry spewing diesel smoke, we bid a reluctant farewell to Camp Hadar. The caravan stopped at Eloaha, where our local workers lived, and after a couple of cold Cokes and heartfelt hugging, we said good-bye and drove off in a cloud of dust.

Roaring along at about 100 kilometers an hour, I was hopeful that we could make it to Addis by early evening, but half an hour outside of Awash Station I heard an explosion, and my Land Cruiser began to fishtail out of control. As I struggled to keep it from flipping over, I heard Bob shout, "Holy shit!" At last I was able to steer into a skid and the car came to a hot, smoky stop. "Close one!" I exclaimed.

A few more flat tires delayed us, and I knew we'd have to overnight somewhere. Around seven we rolled into the rundown Buffet d'Aouache, a rest stop along the Addis-Djibouti railroad that had lost every glimmer of its former glory but at least promised beds and food. After recovering from our disappointment when we learned that they had run out of beer, we dined on veal cutlet and salad, then went to our rooms. I took a cold shower, then attempted to read, but after a few pages of *Jurassic Park* I couldn't keep my eyes open. I turned the lights out at nine.

I was awakened several times by barking dogs, but I had a pretty good night's sleep, and we were back on the road by 6:00 a.m. Before long we had reached the extreme southern point of the Afar Triangle. Bob, with his head hanging out of the Land Cruiser, was as happy as a dog on a road trip; he pointed out fascinating geological aspects of the landscape, including a field of dark lava that showed recent volcanic activity. I enjoyed his discourse on the pristine volcanic cones as we passed the seemingly endless piles of volcanic cinder.

The National Museum of Ethiopia was open when we drove into the compound on Sunday; one of the employees arranged for us to unpack our vehicles and secure the several boxes of fossils we had collected at Hadar. After dropping off our Addis-based crew, we checked into the Ethiopia Hotel. I ordered a chicken sandwich with french fries from room service; after eating I took a nap.

On Monday morning we confirmed our seats on the Lufthansa night flight to Frankfurt the following Saturday. Bob, Bill, and I then went to the museum to officially turn over our fossils to Berhane Asfaw. The whole compound was swarming with activity because Tim White's expedition had also returned, and everyone was reorganizing and storing their field equipment in the garages.

In the afternoon Bill and I met for an hour with Tadessa Terfa at the Ministry of Culture and Sports Affairs to review our field results.

"Very good," Terfa said. "Berhane is at your disposal, and you must follow all of his instructions at the museum." We thanked him, shook his hand, and said good-bye. There was still work to be done on the fossil hominids—measurements, observations, and photographs—and we were assured that this would be no problem at the paleoanthropology laboratory, a modest but well-equipped lab that had been built mostly with National Science Foundation money and private contributions.

Six days passed in the blink of an eye. Bob and Paul had to repack all the geological samples, label everything, print out several copies of the master list of the samples collected, and obtain an export permit from the Ministry of Mines. Bill and I worked side by side as we made important notes on each hominid specimen and took both color and black-and-white photographs. It had been some time since we had worked together on new Ethiopian fossils, and I appreciated his hard work, expertise, and easygoing personality.

On Saturday, our final day in Addis, Berhane walked over to me when I was photographing a mandible and said, "I assume you have made copies of the video you shot in the field for the ministry."

"I'm planning on doing that as soon as I get back to Berkeley and will send them to you," I replied.

"According to the antiquity regulations, they have to be copied before you leave, or you cannot export them."

I stared. "Well, I'm not sure I'll have time today, but I'll try."

As soon as he left, I flew to the room where the video equipment was kept. The VCR kept shutting itself off, and I was never completely sure what I was doing wrong. After several hours I had to give up; there were just too many other things I had to do. Reluctantly I presented myself in Berhane's office to explain. "I'm not sure what I should do now," I said.

He smiled and told me not to worry. He would come to the airport with me to smooth over any problems connected with taking the videotapes and the geological samples out of the country.

Berhane asked me to dupe the tapes when I got home and immediately send copies to him.

I was apprehensive about this arrangement. Berhane was always quick to remind me that any failure to follow ministry guidelines might result in cancellation of the Hadar Research Project. I couldn't help wondering why he was being so accommodating now.

Bill and I left the museum to return to the Ethiopia Hotel for final packing and preparation for our late-night flight. Berhane accompanied us to the airport; no one questioned the tapes or samples we were bringing home.

The Lufthansa flight landed in Frankfurt, after a stop in Jidda, Saudi Arabia, around eight thirty in the morning. Bill, Bob, Paul, and I were traveling together, and after a few hours' layover in Frankfurt, we boarded a flight to San Francisco, where our wives and friends picked us up and drove us to IHO. A large banner strung across the balcony over the parking lot greeted us: WELCOME HOME IHO TEAM! We toasted with a glass of champagne and then went our separate ways to celebrate homecoming and the holidays.

I was eager to get to the office after having been away from IHO for such a long time, so I went in the day after Christmas. My secretary, Larissa Smith, looked upset. "There are two faxes on your desk that won't make you very happy."

With a sinking heart, I recognized the official letterhead of the People's Democratic Republic of Ethiopia, Ministry of Culture and Sports Affairs. Both faxes had been written on December 24 and were signed by Tadessa Terfa. One read, "Video records should not have been exported without review and formal approval!" The other objected to the exportation of the geological samples "without proper arrangement of permits between the Ministry of Culture and the Ministry of Mines."

I called Bill to give him the news.

"What the hell? You said Berhane gave us the go-ahead!"

"He did!"

Bill didn't have any idea what was going on, and neither did I. All I knew was that future work at Hadar could be jeopardized.

"We'll just have to hope that the ministry will accept our apologies and understand that it was never our intent to violate antiquities regulations," I said grimly.

Several Successful Field Seasons

With the rousing success of the 1990 Hadar field season behind us, my thoughts turned to planning for our next trip. We returned the videotapes and apologized profusely for the misunderstanding over the protocol for the geological samples, which had satisfied the ministry officials in Addis. But I knew that going back in the fall of 1991 was still pretty iffy. The political situation in Ethiopia was heating up. Mengistu's troops were being routed everywhere by a coalition of the Ethiopian People's Revolutionary Democratic Front and the Tigrayan People's Liberation Front. These well-seasoned fighters were determined to liberate Ethiopia from a brutal dictatorship. Five months after I left Ethiopia, they surrounded Addis Ababa. Mengistu's time had run out. He fled to Zimbabwe on May 21, 1991, and on May 28, Meles Zenawi, a Tigrayan who is now prime minister of Ethiopia, marched into the capital and took control. Marxism was gone and in its place was a fledgling democracy.

In the immediate aftermath of the liberation, Ethiopia was largely in chaos, so I wasn't too surprised when Tadessa Terfa's June 17 fax arrived: "Currently we are on transition toward a new government . . . we advise you to reconfirm with us prior to coming to Addis Ababa for fieldwork."

"Think we'll get to go?" Bill Kimbel asked.

"I don't know." I sighed, wishing that the fossil-rich region was located in a peaceful, remote section of Vermont or Colorado! "Let's hope for the best."

I hung on to a thread of hope, but a shiver went up my spine when another fax from Tadessa arrived in mid-September: "Since conditions are not conducive for work this fall, we are obliged to postpone your field permit."

"Postpone? For how long?" Bill demanded. "Weeks? Months? Years?"

This time I didn't answer. I just stared gloomily at the paper. We'd had such an exciting 1990 season. We needed to get back into the field. We needed to find an *afarensis* skull.

Unbelievably, later that year we received notice that expeditions were encouraged to plan for fieldwork as early as the beginning of 1992.

"Finally, some good news!" Bill said.

"Let's get going right away," Bob Walter suggested. "Don, how fast can you hustle up some funding?"

"I'll start making calls first thing tomorrow," I promised. "In the meantime, let's get a team organized." That was the easy part; there were lots of experienced colleagues who would jump at the chance to join us. For a few hours we proposed candidates, and Bill said he would coordinate the paperwork. We were going back, I rejoiced. And I had a feeling we'd find a skull this time. We had to.

Bill left for Addis Ababa in January 1992 and with Tadessa's support he led a reconnaissance to the Afar region to assess the security situation. When I arrived in Ethiopia in mid-February, he told me that he had a successful visit with the recognized leader of the Afar people, Sultan Ali Mira, who encouraged us to return to Hadar and assured us that the area was peaceful.

That was good news, but what concerned me was our reception at the Ministry of Culture and Sports Affairs. I'd been so anxious about the two letters we received in Berkeley, but when I reported to Tadessa's office, I was greeted with a warm smile. "We want you to know that if there's anything you need, you just let us know. And if you notice anything out of the ordinary, please report it to us, and we'll handle it." At the National Museum, Berhane Asfaw was also friendly and respectful, as he welcomed us back and told us to let him know what he could do to help.

The country was clearly in transition from Mengistu's regime to the new government, and the atmosphere in Addis Ababa reflected the turmoil, uncertainty, and lingering fear. The roads were considered unsafe and the news was filled with reports of truck drivers who had been attacked, so our caravan had to be escorted north by a large military convoy. Being locked in on all sides by heavily armed soldiers and pickups with 50-caliber machine guns was depressing and a little scary. I knew we didn't face any real danger in the field, but after the troops left us at Odaitu, I was relieved when the Afar Regional State government assigned five Afar Liberation Front troops to look after us.

"A hundred and eighty degrees from the way they treated us last time, huh, Bob?" I commented.

"You gotta wonder why."

Unable to come up with a single explanation that made sense, I simply shrugged. "This is one of those things we just can't worry about right now. Agreed?"

"Agreed."

Once installed at camp, my excitement shifted into high gear. "Can you believe we're here? Something tells me this will be a great season!"

"It sure got off to a positive start," Bill agreed. He was referring to the news that greeted us when we got there. During our absence, Dato Ahmedu, one of our distinguished fossil finders, had already spotted a hominid mandible fragment lying in the sand during a walk through Hadar, and he took us directly to where he had left a cairn. That's the thing about Hadar, I thought for the millionth time. One minute you're out for a walk, the next you're picking up a crusty brown specimen, another hominid fossil. That it could happen as fast as it takes your vision to focus still amazed me.

I was also very excited about the field season because a film crew would be joining us at Hadar. Earlier in the year, during a trip to Cairo, I'd met a science editor for the program *NOVA* who was interested in working on a project for PBS with me as the narrator and host. We came up with the idea to shoot a three-hour series, *In Search of Human Origins,* and the first episode, entitled "The Story of Lucy," would be filmed at Hadar.

Another focus of the research was on the archaeological work in the Gona region. Jack Harris joined our group to supervise his student Sileshi Semaw, who would begin the first stage of his Ph.D. work. The stone tools at Gona were thought to be more than 2.5 million years old, and expanded excavation promised to provide us with insights into early hominid behavior.

As for Bill and me, we'd decided to concentrate our survey efforts on more recent geological horizons of the Hadar Formation from which no fossil hominids had yet been recovered. These strata were geologically younger than Lucy, and if they yielded hominids, we could catch a glimpse of what Lucy's descendants looked like. After careful scrutiny of the geological map and consultations with Bob Walter, we chose an area just north of the main Kada Hadar drainage whose landscape of powdery clay sediments earned it the name Buruki, the Afar word for "soft soil."

I enjoyed walking this area. Every once in a while I'd summit one of the higher knolls in search of a breeze. From that higher vantage point, I could see which slopes to explore next. Then I would surf down the soft slopes of clay and continue my survey.

On the fourth day of the season I spotted a large white fossil high on a sandy outcrop, which, as I drew closer, I recognized as a partial lower jaw of an elephant. From there I took a moment to make my decision: Turn right or left? My gut said Turn right, which I did, and I began to follow a sandstone layer that was running alongside a small gully. With my eyes glued to the ground and hearing only the crunch of the sediments as I slowly made my way, I suddenly spied something that made me stop dead in my tracks.

Nestled in dry, yellow grass was a proximal ulna, an elbow bone just like the first bone I found of Lucy's skeleton. But when I crouched and gently eased it out of the sand, I saw that it was twice the size of hers. "Has to be a male!" I said out loud. I made a quick scan of the immediate area—nothing. I built a little cairn of stones to mark the location, then punctuated it by plunging my rock hammer into the sand. I knew Bill was nearby, and with the ulna securely stored in a cloth bag I strode off to find him.

He looked up at the sound of my approach, then stood when he saw the look on my face. "Find something?"

"Look!" I presented the ulna.

"You think hominid?"

"Yes!"

"You sure? It's really big."

"Hominid," I insisted. "I want to bring it back to camp and compare it to Lucy."

"Before you go, I want to show you a couple of hominid frags we found during screening," Bill said.

"I'll be right back," I promised. I needed to get my hammer and mark the locality, A.L. 438. Just as I finished marking the spot with a yellow survey flag, I heard footsteps behind me and saw Bill and Yoel Rak approaching. Apparently Bill had told Yoel about my find.

I stood up and handed Yoel the ulna. "Well?"

"No question about it—hominid!" he replied, looking around. "Any other fragments?"

"Unfortunately not, but I haven't really done much of a survey here." Just then, I spied a small pile of white bones lying on top of a crest about 150 feet away. We raced up to the spot.

"Well, look at this!" I exclaimed. There were a couple of meta-carpal bones from the back of the hand, a fragment of upper arm bone, or humerus, and a fragment of skull.

Bill crouched and lifted out the skull fragment. "Beautiful! Perfect! It's the glabella!"

This was real cause for celebration, because the glabella—the area just above the nasal opening between the eye sockets—was one area of the adult skull unknown in *A. afarensis.* Finding a skull fragment is always exciting, but to find a never-before-discovered bit of anatomy is as great as the difference between seeing a drawing of a triangle and actually standing in the dim, chilly interior of the Great Pyramid of Giza.

Bill scrutinized the glabella (Latin for "bald") and pointed out the prominent ridges where the superciliary muscles attached, the ones that draw the eyebrows down and closer together. I joked, "He would have been able to make quite a frown, eh?"

Frowning a little himself, Bill said, "Better get an excavation team organized and see what we can find in situ."

"Let's get on it right away," I agreed happily.

However, after two days we were forced to put the plan on hold. Bob and I had been out to visit Jack and Sileshi in the Gona area, where their excavation was unveiling numerous artifacts. When we drove into Camp Hadar later that afternoon, we hadn't even shut off the engine before Bill came running out to the car.

"The hits keep on coming."

Bob and I jumped out and followed Bill to the worktable where Yoel sat grinning before a dozen skull fragments, including parts of the face and braincase. The bones were large and ruggedly built, unmistakably male.

"I found it!" he said.

I was thrilled for Yoel because during the 1990 field season he had not found even the tiniest fragment of hominid, much to his frustration.

"So I see!" In awe, I pulled up a chair and looked over the collection before asking, "Which bone did you find first?"

"This." He picked up a fragment of occipital bone, the rear portion of a skull, and handed it to me.

The bone had turned up just off the main Kada Hadar near the Buruki drainage. Still grinning with pride, Yoel explained that other skull fragments were clustered around the occipital and because the skull sutures were fused, he was certain they all belonged to an adult *afarensis* skull. From the large size and heavy muscle markings he concluded it must have been a male.

"Son of a gun," I said. "Nice work!"

At last we had a skull. After taking in the enormity of Yoel's discovery, I wandered over to the showers, where I ran into a squadron of mosquitoes; I quickly returned to my tent. Another remarkable day at Hadar. How many other hominid fossils were out there, each waiting to reveal its own story? As we sat down to eat dinner, I raised my glass to toast Yoel and his fantastic discovery.

By the time the *NOVA* film crew arrived three weeks later, we had collected hundreds of bone fragments representing 75 to 80 percent of the skull Yoel found at A.L. 444. Unfortunately, Yoel had to return to Israel for university duties, so Bill explained the skull for the NOVA team. He pointed out that the A.L. 444 specimen was one of the largest and most heavily built *Australopithecus* skulls ever recovered, and was most certainly from a male. Slightly bigger than the skull of a female gorilla, it showed strong resemblances to the African apes, with its massive, projecting face and relatively small braincase. The most important thing about 444 is that it was the first specimen to preserve the braincase and the lower jaw of a single adult *A. afarensis*.

When work resumed at the A.L. 438 site, the camera was rolling as Bill and I excavated more of the ulna. Back in camp the film crew captured my delighted reaction as I cleaned the fragments and fit the broken pieces back together. "A perfect ulna," I declared.

The ulna, along with the other bones we had retrieved from the site, also provided the first definite association between hand bones and an upper limb for any species of *Australopithecus*. (Lucy's skeleton, as well preserved as she was, only included a single wrist bone and one segment of a finger.) Why is that so important? a member of the film crew wanted to know.

"The relationship between hand bones and arm bones can tell us whether *A. afarensis* did much climbing," I answered. Referring to the fossils on the table before me, I explained that the ulna and the fragment of humerus showed well-developed muscle attachments that might help to sort out upper-limb mechanics. And the male ulna was twice the size of Lucy's. "That's significant," I said, "because now we can compare the anatomy of these two specimens to see if they're identical or distinct. In other words, it might help settle the question of whether there were two species here or one."

Each day was spent toiling for hours under the blistering Ethiopian sun. During Ramadan our Afar workers, in particular, suffered, since they were forbidden to eat or drink from sunup until sundown. A hundred times I wanted to warn them about the dangers of dehydration. Their beliefs had a much stronger hold on them than any advice I could dispense, however, no matter how logical it seemed to my Western sensibility.

On the third day of Ramadan we were engrossed in work when suddenly we heard the explosion of a gunshot. Even before fear registered, I jerked up my head and swiftly inventoried each face—the only one missing was Omar Abdulla, our camp entertainer, who was known for his uncanny imitations of us. He'd been nearby just moments ago. Scared, I hurried off toward the sound of the gunshot, hoping he was okay.

To my relief, I found him standing by the watercooler, apparently unharmed. *"Mendano?"* I asked. What happened?

He was shaking all over and pointed to a snake half shattered by his bullet. *"Mamehe!"*

"Yes, I see it's bad, but what happened?"

He pointed to the sky. "Allah!"

In a moment I figured it out: He'd snuck off to get a drink, had been startled by a snake hiding under the cooler, and shot it. Obviously a sign from Allah that Omar should not violate the laws of Ramadan!

The animal encounters didn't end there. A few evenings later two mangy old lions began to circle camp, grunting and roaring. We learned that they had killed a donkey nearby, and now seemed to be looking for dessert. Their throaty snarls during dinner unnerved us, and even drove some team members like Jack to sleep in the safety of a Land Cruiser. I braved the night in my little tent, hoping the strategically placed fires and the Afar guards would keep the lions away, but when I awoke in the morning, I unzipped my flap and saw large lion paw prints right outside. When I told Bill, he commented that I must not have smelled good enough to eat.

The 1992 field season ended, and we packed up and headed for Addis Ababa. At the entrance to the National Museum of Ethiopia we were met by a policeman who denied us passage to the back where the storage sheds and the paleoanthropology lab were situated. I tried to explain that we needed access to the sheds and the labs, but he kept shaking his head. *"Yelem,"* he said. No.

As we debated what to do next, to our surprise Sileshi ran over to our car; he reported that a ministry grievance committee had investigated Berhane and Tadessa and both of them had been dismissed from their posts at the ministry. Furthermore, Berhane had sealed the lab before he departed for the United States, so there was no way for us to get in.

I turned and stared at Bill. We had no choice but to drop off the fossils at the museum and wait. Thankfully, within just a few days a new museum director and a new head of the Centre for Research and Conservation of Cultural Heritage were appointed, and we were relieved to move the precious fossils into secure safes inside the lab. We were also able to retrieve our personal gear from the sheds. The administrative transition made it difficult to accomplish much work in the museum. But at least we had found our hominid skull.

Season after season Hadar never disappointed. Our storehouse of hominids continued to grow and grow. I knew that each and every fossil found, whether it was a jaw fragment with a few teeth, a bit of limb bone, or a skull like A.L. 444, enlarged our understanding of Lucy's species. My goal was to make *Australopithecus afarensis* the single best known hominid species from anywhere in Africa.

To accomplish this goal, we designed and refined a rigorous and strictly enforced protocol: A Hadar Hominid Specimen Record Sheet had to be filled out by a paleoanthropologist to record each detail about every fossil find. A locality number and a specimen number are assigned. The date of discovery and the name of the finder are recorded, and a brief anatomical description is provided. Next, the form is given to a geologist who identifies the exact position of the find in the Hadar Formation, indicating whether it was buried, half buried, or lying on the surface. The conditions of the discovery are recorded, and photographs are taken of the specimen as it was found on the ground, and of the surrounding area. Using a GPS device, the exact coordinates are determined and recorded. The specimen is removed, put into a labeled plastic or cloth bag, and brought to camp, where its specimen number is entered into the master faunal catalog.

It never occurred to me that success at Hadar could have a downside, but it did. With the burgeoning geological work and demanding logistical duties, work at A.L. 333, the First Family site, kept getting put off. Each season we scheduled brief side trips there and always recovered a few hominid fragments, but we needed to undertake a more comprehensive exploration. A major challenge we faced in 1994 was that after nearly two decades of erosion, virtually all traces of our earlier work had been obliterated. We needed to move tons of newly eroded sediments, screen them on the off chance that they contained fossils, and then conduct large-scale, backbreaking excavation. The only way to tackle this knotty problem was to find someone who knew excavation and didn't shy away from hard work. Yoel recommended an Israeli colleague named Erella Hovers, who he said would be a real asset because she was not only an accomplished excavator but also a distinguished archaeologist. So Erella joined us for the 1994 field season. High-strung and indefatigable, she was everything Yoel promised. Yet after moving tons of sediments at A.L. 333 and finding twenty-six hominid fossils, Erella's excavations did not turn up a single fragment in situ. The precise level from which we'd discovered fossils in the 1970s was located, but after arduous excavation, Erella concluded that the horizon was exhausted.

"Great work, and even though it's disappointing that we didn't make any remarkable finds, knowing that the in situ horizon is no longer promising allows us to put our efforts into something else," I said that evening over dinner.

"All information is good information," she agreed.

"But it still leaves me with one question . . ."

"You want to know how they died," she anticipated.

"Thirteen individuals all at once. Maurice Taieb suggested it was a flash flood. So what do you think?"

Erella shrugged. "I really don't know. But if you find out, let me know."

Over the next few field seasons, we made modest additions to our collection of fossils, and we continued to study the remains intensively, fleshing out our portrait of *A. afarensis* in ever greater detail. But it wasn't until 2000 that the team made another landmark fossil find. Urgent IHO duties back in the United States forced me to make an early departure from the sixth Hadar field season in 2000.

On October 26 I headed south to Addis Ababa. The 600-kilometer drive to the Ethiopia Hotel was uneventful, and after a quick shower, I called room service to order pasta and a bottle of Meta beer. As soon as I hung up the phone it rang. It was Gerry Eck.

"Everything okay?" I asked.

"You bet! Dato Adan found a skull this afternoon out in the Unda Hadar ["little Hadar"] drainage area!"

"You've gotta be kidding! What's it look like?"

"Fossilization is excellent and it doesn't appear to have suffered much deformation. Most of the mandible, face, and braincase look preserved."

"Just my luck, the day I leave the field! Male or female—can you tell?"

"Hard to say. It isn't massive like 444, but it's definitely not as small as Lucy."

The skull, which would come to be known as 822, sounded like a female to me. I could only imagine how excited Dato was. He had wanted so much to add a skull to his long list of discoveries. And now he had. "Great news, Gerry, thanks for letting me know. Man, I wish I could have been there!"

"You would have loved it, Don. Dato was thrilled."

"I bet. Tell him I said *kada mehe*! Very good!"

"Will do. Okay, gotta go. Have a safe trip. Signing off."

The satellite phone went dead and I felt a million miles away from the action. But I consoled myself by thinking about how finally, twenty-six years after finding Lucy, we would be able to see the face of a female *A. afarensis*.

With so much else going on, we'd been derailed in our efforts to determine what had caused the death of all the individuals at the 333 site. But we had never completely given up hope of finding the answer. Bill suggested that we invite Anna Kay Behrensmeyer, known to everyone as Kay, to mull over this problem. An unassuming, soft-spoken woman who gives the impression that she would be more comfortable knitting a sweater than scrambling over rocky outcrops in Africa's Great Rift Valley, she's an experienced geologist who has spent more seasons in the searing African sun than I have. With her expertise in taphonomy—the study of the processes that influence bone from the time of death to the day of discovery—she was uniquely positioned to figure out the circumstances that had produced the unique collection of hominids from A.L. 333. We contacted her at the Smithsonian Institution, and she was thrilled to accept our invitation to help solve the conundrum of the First Family.

The 333 locality sits in the middle of a gigantic amphitheater of steep sediments capped by a thick resistant sandstone. Kay spent part of the 2001 season there, and near the end of her stay I drove out to see what, if anything, she'd turned up.

"I've been all over the hillside," she told me. Her sleeves were rolled up and her forehead was glistening with sweat. She and other members of the team had dug seven 8-meter-high geological trenches lateral to where the hominids were found in order to figure out how the deposits accumulated.

"So, what was the environment like here when the hominids were buried?" I asked.

She opened her metal clipboard and consulted handwritten notes. What we were looking at here, she explained, was a paleo-channel. A 30-meter-wide, fairly shallow river had cut through this area, and the hominids were preserved smack-dab in the middle of the channel.

I looked at her graph paper where she had even outlined the course of the channel. "So, it was a flash flood?" I asked.

"Nope." She showed me on her sketch the precise level of the geological section that originally contained the fossil hominids, noting that the sediments there were very fine grained, essentially a silt that was deposited in an abandoned channel swale.

"How sure can you be about that?" I inquired.

Kay replied that the cross-bedding suggested rapid deposition, indicating a very gentle flow of water that was traveling 50 degrees east. If the water had been a raging flood, she said, the many small and delicate bones we found there—including hand and foot bones, along with ribs and vertebrae—would have been swept away. She added that an important taphonomic clue was that all the bones show a narrow range of weathering, meaning they had all been out on the surface for roughly the same length of time before getting covered in silt. Another clue, she said, was the fact that the weathering was not very intense, so they were interred quickly—perhaps in a matter of weeks. Although Kay was convinced that the 333 hominids were part of a group that probably foraged together, she was more cautious about whether they all died at the same time. Their interment was probably a single event, but the deaths of the individuals may have transpired over a more prolonged period of time.

"Well, so, what killed them?"

"I have to say . . ." She paused.

"Yes?"

"I don't know. Pretty much all I can say is that it wasn't a flood. A few of the bones I examined in Addis show signs of carnivore damage, and my guess is that some scavenging occurred before final interment."

"Hyenas?"

She took a gulp from her canteen, then offered it to me, but I shook my head. She thought the carnivore was some sort of a cat. Previous research had shown that when cheetahs kill baboons, they leave the hands and feet behind.

It occurred to me that we had found cheetahlike creatures, including saber-toothed cats and even leopards, in the Hadar deposits. Maybe cats had chased some of the First Family into this area of mud or quicksand, where they became mired.

How about that? The account was so different from what I had expected. I was sure she would confirm the flash-flood theory. But one of the most important lessons I have learned is that it's vital to be open to explanations that you never considered before. Often the theories that take you most by surprise turn out to be correct.

Each evening the Afar kneel within a circle of rocks and dry grass to recite a prayer, their soft murmurs hypnotic. Respectful of the ceremony, I always granted them privacy. But one evening during the 2003 field season, I had to pass by them on my way to my tent. Moving as quietly as I could, I observed the group bowing in unison. Then, to my surprise, I heard three familiar words spoken: *Johanson, Allah, Lucy.*

I was reluctant to seem nosy, but curiosity prevailed over grace, and I asked our representative from the Afar Regional State government, Mohamed Ahamadin, if he knew what they had been saying. He promptly motioned Omar over and engaged him in quiet conversation that I couldn't follow. Then he turned to me and explained, "The Afar were thanking Allah for bringing you here to Hadar so that you could find Lucy. They ask each night that Allah keep you safe and bring you back to Hadar soon."

I was touched. "What do they think of Lucy?" I asked. "I know it's hard for them to grasp the concept of millions of years, and I doubt there's much room, if any, in the Koran for the notion that modern people evolved from distant ancestors."

"They believe that Lucy was the first human," Mohamed replied. "Since she was found here in Hadar, she was an Afar. So, for them, all humans on the earth today are descended from the Afar people. And that makes them very happy."

"Well," I said, "that makes me very happy, too."

CHAPTER 6

Getting to Know Lucy Better

H ow many?" I asked, thinking I'd heard wrong.
Three hundred sixty-five.

"How did the number get so high?"

We had added 115 over the past few seasons, Bill Kimbel replied. He tapped the keys on his computer until an extensive list appeared. "Check it out."

I let out an appreciative whistle. Hominid specimens representing 365 individuals from Hadar! For most of 2004 I'd been so caught up in administrative duties related to the institute, traveling and fund-raising, that Bill was in charge of the record keeping. Now, sitting in the "hominid room" at the National Museum of Ethiopia, I was glad to be back on the scene. "Finding the male and female skulls, that was really the icing on the cake, huh?"

In response, Bill commented that Sterkfontein had nothing on us. I knew he was referring to the more than six hundred hominid specimens recovered from the limestone cave site of Sterkfontein in South Africa. It was a mind-boggling number, but unfortunately, vagaries of geology and deposition compromised the integrity of the collection. Over time, sections of the roof had given way, confusing the stratigraphy. Even worse, younger fossils had fallen into the subterranean cave, often rolling down a debris cone, and were buried cheek by jowl with much more ancient fossils, leading to false associations.

Another reason to love Hadar, I thought. Our specimens of a single species had made their way to the surface, then waited patiently until we found them in a precisely calibrated sequence covering nearly half a million years of time, at which point, with the aid of modern technology, we were able to definitively ascertain their ages.

"So where do we start?"

Bill wanted to study the A.L. 444 skull. Like me, he believed that the anatomy of the skull would clinch the distinctiveness of *A. afarensis* from all other *Australopithecus* species. For years the lack of a skull had left us unable to make a comparison and produce a definitive scientific monograph; all we'd been able to do was speculate. But 444 was in about a thousand fragments, some of which had become severely misshapen. Reconstructing it was going to be a daunting task that I wouldn't even begin to have the patience for. Luckily, Bill and Yoel Rak, two of the world's foremost experts on the *Australopithecus* skull, were actually excited about piecing it back together.

It took three months to liberate the skull fragments from a cemented layer of silt and assemble the fifty or so fragments that constituted eight major portions of the face, the braincase, and the lower jaw.

When I saw it, I remarked that the skull appeared to have suffered some serious deformation. Yoel nodded, noting that the major trajectory of the geological forces that deformed the braincase were directed downward from above the right orbit toward the left lower region of the back of the skull. Knowing this would allow us to recognize where to make corrections during our reconstruction. The plan was to glue the pieces together and make a rubber mold of the original. Then Yoel would pour a resin into the mold to make a cast, and before it fully hardened he would bend the deformed areas back into normal anatomical position. It doesn't always fix distortions 100 percent. But it's pretty damn close.

I couldn't wait to compare Yoel's reconstruction of 444 to the composite skull that Bill and Tim White had created back in the mid-1970s based on fragments from twelve adult Hadar specimens. Their major concern at the time was that they had to guess at the shape of the brow and the frontal bones of the vault because Hadar had not yielded any fossils from that region of the skull.

It turns out that 444's frontal region of the skull is more convex and steeper than we theorized. That means the braincase is higher and more rounded. The way the face fits onto the cranium is also different from what Bill and Tim had theorized, tilting downward, which lowered the snout and resulted in a somewhat less projecting, apelike visage. Aside from these minor alterations, Tim and Bill had done a pretty good job.

Pleased, I took Yoel's reconstruction of 444 into my hands and turned it slowly. "Boy, this guy was a real bruiser." It was the largest *Australopithecus* skull ever found—bigger than that of an average female gorilla, Bill and Yoel thought. When Bill took out a pair of calipers and measured the girth of the skull just behind the orbits— the so-called postorbital breadth—it was a whopping 77 millimeters, 25 percent larger than that of any other *Australopithecus* skull. Yet its face is taller and less projecting than an ape's face, and the brow ridges are delicate. It's definitely more hominid than ape.

Yoel suggested we compare it to the female cranium that Dato Adan had found. I picked up the A.L. 822 skull, taking a moment to admire how much better preserved it was than 444. My calipers confirmed that this specimen, too, was very broad in the postorbital region—more evidence for the distinctiveness of *A. afarensis.* Although the 822 specimen is much smaller than 444 in its overall proportions, the two are identical in shape. "Nothing suggests these were two different species at Hadar," I pronounced.

* * *

The question of how many species are represented in the hominid fossil record is one of the thorniest topics in paleoanthropology. During much of the early part of the last century, scientists commonly named a new species for virtually every fossil specimen discovered. In contrast to these scholars, dubbed "splitters," there arose a group of "lumpers," who emphasized the fact that there are varying levels of anatomical variation within modern primate species. The lumpers argued that this variation in modern primates must be taken into consideration when looking at fossil specimens. Viewed that way, they contended, the number of fossil hominid species would be reduced. This point was well taken, but in the early 1950s lumpers pared so many species from the human family tree that the true diversity of hominid evolution was lost. Paleoanthropologists today incorporate a more comprehensive understanding of the process of evolution and speciation.

Every meeting of paleoanthropologists that I have attended has at some point erupted into intense and sometimes unfriendly debate about how many hominid species preceded us. Some evolutionary biologists contend that a species is an abstract concept that we impose on nature. But for the evolutionary biologist Ernst Mayr, who lived to be one hundred years old and wrote more on the nature of species than practically anyone, species were actual entities in nature, which he considered "the keystone of evolution." Species are the units of biological change, and it is therefore of utmost importance for understanding human evolution that we correctly identify, as best we can, the number of species on our family tree.

Traditionally scientists have based the definition of a species on living organisms. Mayr's biological species concept, as it is called, says that a species is a group of organisms that can successfully breed with one another and cannot exchange genes with other organisms. To punctuate this point, when I teach about species in my introductory human origins class I wear a button that reads, "Sorry, I don't date outside my species." Take, for example, our household pets.

The family dog is *Canis familiaris* and the family cat is *Felis domesticus*. They look different, act different, and don't interbreed to produce offspring because they are biologically and reproductively isolated from each other.

What do we do when a species is extinct and unavailable for real-time scrutiny? Obviously, we can never say for certain whether in the distant past the two fossils sitting on my lab table were capable of reproducing and thus constitute a species. We faced such a challenge when we addressed the biological meaning of the Hadar and Laetoli fossil hominid assemblages. How did we arrive at the conclusion that all of the these specimens belong to a single species? First, we carefully looked at the detailed anatomy of the fossils and established that the lack of a large brain, the presence of a projecting face, features of the teeth and jaws, and so on, were shared with a variety of species already placed in the genus *Australopithecus*. We then looked at the range of variation in anatomy and size within the Hadar and Laetoli samples and compared this to the degree of differences found in living species of primates, particularly the African apes, our closest evolutionary cousins. Once we determined that the amount of variation in the fossil sample did not exceed that seen in a species of the extant apes we felt confident that the hominids constituted a single species of *Australopithecus*.

We then compared, one by one, features of the teeth and bones of the Hadar and Laetoli collections with other known species of *Australopithecus*. What emerged from this comparative exercise was a list of distinctive anatomical traits in the fossil samples that were discrete from all other species, especially *A. africanus*. Based on this exercise we were convinced that the Hadar and Laetoli collections belonged to a distinct species, *A. afarensis*.

This conclusion, based on a comparative study of the fossil material with living and other extinct species, implies that all *A. afarensis* individuals, based on mate recognition, would have been able to interbreed, exchange genes, and produce offspring. This is the logical implication of placing them all in the same species, a supposition that we can never absolutely prove, because these creatures are no longer living, but one that is consistent with all we know about the nature of a species.

Identifying a species, especially in the fossil record, can be tricky. My close friend and colleague Ian Tattersall, of the American Museum of Natural History in New York City, points out that many living species, such as the lemurs of Madagascar, have virtually identical skeletal anatomy but differ remarkably in their behavior, fur color, markings, and even mating calls. Individuals in these different lemur species do not reproduce with one another in the wild. "Were we to step a million years into the future and study fossils of lemurs and base our assessment on only the skeletal remains," he says, "we would greatly underestimate the number of species."

I am not a "lumper" or a "splitter" in any traditional sense, but I do take Ian's suggestion to heart. After an exhaustive study of variation in fossil hominids, Ian and his colleague Jeffrey Schwartz, from the University of Pittsburgh, postulated many more hominid species than are embraced by the majority of paleoanthropologists. Although I believe they have overestimated the number of fossil hominid species, I understand their perspective and am certain that as larger fossil samples are recovered, new species will be recognized. For now, I prefer to look at fossil hominid variation within the analytical framework we developed for determining the distinctiveness of *A. afarensis*.

One day I was discussing with Bill the subject of species, one of his specialties, and I asked him whether, with such a large sample of *A. afarensis* at Hadar, we should see some changes within that species over the 500,000 years. Yes, he said, noting that a species that persists over time—what we call a "phylogenetic species," like *A. afarensis*—is a likely candidate for documenting such anagenetic change. To do this, Bill said, we needed to assess the large sample of teeth and mandibles we had amassed, to see if there are changes in anatomy or size over time, from the most ancient layers at Hadar to the most recent. The work was going to require some sophisticated statistical analyses, so we asked Charlie Lockwood whether he was interested in working on this problem.

Charlie was a taciturn and very bright young scholar who had been a postdoc at IHO. He died tragically in a motorcycle accident in 2008. He was not just a terrific anthropologist but also a math whiz who had distinguished himself at the institute by applying a variety of quantitative methods, predominantly statistics, to evaluate a host of questions concerning early hominids. He'd also recovered a nifty mandible at Hadar, A.L. 729, which was found in the most recent levels of the Hadar Formation, making it one of the youngest *A. afarensis* specimens known at Hadar.

The first thing we did was take simple height and breadth measurements of the body of all *A. afarensis* mandibles to see how they clustered, relative to a sample of modern apes, in our case, gorillas. The pattern in *A. afarensis* and gorillas is essentially identical. The mandibles fell into three clusters: small, medium, and large. For the gorillas, all the large mandibles belong to males, all the small ones to females, and the intermediate group is where large females and small males overlap. Because the pattern of distribution is identical in both gorillas and *A. afarensis,* this helped corroborate our hypothesis that the hominids at Hadar made up a single species, with the large individuals being males and the small ones being females.

Charlie further determined that when all the mandibles from the entire 500,000-year time span were plotted, the degree of variation between large and small exceeded that of all comparative samples for living species of apes. The really large mandibles came from the more recent levels and the smaller ones came from the older levels. It seemed we were looking at a bona fide example of anagenetic change. Bolstering this interpretation was the fact that the mandible shape and anatomy stayed the same across time.

"Meaning?" I knew the answer to this but had to hear Charlie say it.

"Meaning that there is a single hominid species at Hadar, *afarensis*."

The work with Charlie underscored one of the inadequacies of using only modern ape collections for determining whether a collection of fossil *Australopithecus* constitutes one or more species. Modern ape collections sample living primates, not chimp and gorilla species over time, so they convey variation in contemporary individuals, but we have no idea of anagenetic change over long time periods because the fossil record for chimps and gorillas is virtually non-existent. That is precisely why the Hadar collection, I believe, will become the type collection for assessing all other species of our earliest ancestors.

For advice on the postcranial remains—the bones below the neck—Bill and I went directly to Carol Ward, a highly respected paleoanthropologist, at the University of Missouri–Columbia. Carol's expertise even before she left graduate school at Johns Hopkins University translated to impressive publication credentials. But what I like most about her is her enthusiasm. Fresh-faced and energetic, she is the kind of scientist who is constantly busy, but always finds time for one more thing. Sure enough, she couldn't resist the offer and readily agreed to work on the *A. afarensis* specimens.

"Isn't he something!" she said, admiring the large partial male skeleton I found in 1992, the one with the enormous ulna that's now cataloged as A.L. 438. Carol suggested we let her graduate student Michelle Drapeau analyze the partial skeleton for her Ph.D. dissertation. Ever since my Cleveland days I have known that giving young, eager scholars access to original fossils was a sure way to jump-start their careers. Without hesitation I answered, "Why not?"

Michelle, a petite, dark-haired Canadian who is now at the University of Montreal, turned out to be a delight and a great help in the field. She was as thorough a researcher as I had ever met and she coaxed more information out of the fossils than I could have

dreamed possible. Bill and I were most interested in hearing what
A.L. 438 said about the degree to which *A. afarensis* might be arboreal. We were 100 percent convinced that *A. afarensis* was fully
bipedal on the ground, based on the anatomy of the hip, knee, and
foot. But with the discovery of 438's arms, we wanted to know if
A. afarensis also spent time in the trees.

Carol and Michelle had intensively studied forelimb proportions, which can reveal a lot about locomotor behavior. Michelle
noted that 438's metacarpals are nowhere as elongated as those of
tree-climbing chimps. And although the ulna is relatively long compared to yours or mine, it's significantly less elongated than the ulna
of an orangutan, which, of course, is almost exclusively arboreal. If
A. afarensis had spent a lot of time in trees, there should be a
stronger signal in the forelimb reflecting that behavior.

"What did the brachial index show?" I asked, referring to the
method of dividing ulna length by humerus length and multiplying
that figure by 100. A relatively long forearm, such as that seen in
chimps, will clock in at 95 percent. A relatively long humerus, as
seen in modern man, gives an index closer to 80 percent. Lucy's
ratio was comparable to a chimp's at around 92, so we expected to
see something in that range. The problem was, 438 was missing most
of its humerus. So Michelle and Carol used the humerus Tim White
had found in 1990 (known as A.L. 137-50), which was similar in size
and robustness. The resulting brachial index turned up a ratio of 91
percent, confirming that arm proportions are the same in large and
small *A. afarensis*. The relatively longer ulnas in *A. afarensis* further
suggested that natural selection had not yet shortened the forearms
inherited from a more primitive ancestor, Carol added.

"What about the metacarpals?" Bill asked.

Carol explained that compared to chimps, whose elongated
metacarpals facilitate knuckle walking and moving around in the
trees, and gibbons, which employ their long hands and powerful gripping abilities in arboreal acrobatics, *A. afarensis* possesses

short metacarpals. Indeed, they are even shorter than those seen in modern humans, meaning they weren't exceptionally well adapted for tree dwelling.

Neither did 438 hint at the existence of a second species at Hadar. Michelle reported that the ulna, radius, and humerus of 438 and 137 show no appreciable differences in anatomy, joint orientation, or patterns of muscle attachment from Lucy's. "We concluded that what we're looking at are predictable size variations within a single species," Michelle said. In other words, females like Lucy are small, and males like 438 and 137 are large.

Carol observed that when we only had the tiny bones of Lucy and much larger bones from other Hadar localities such as 333, 438, and so on, there was a significant size gap between small and large specimens. Since the Hadar collection now included intermediate-size specimens, we could begin to see a continual size variation from small to large, characteristic of a single species.

Michelle's analysis couldn't have rendered better results. But what were these large males doing with such big arms? I wondered. Carol remarked that the orientation of the olecranon process—the elbow—of 438's ulna and the notch suggests that this joint was optimized for loading in a 90-degree position, more suited for carrying something, not for hanging from tree limbs. Again, no evidence for climbing behavior.

Carol's work on vertebrae additionally suggested that Lucy had a condition known as Scheuermann's disease that isn't found in any primates other than hominids. She suspected it was caused by the bending of the vertebrae that occurs in association with exclusively bipedal walking. So we had yet another piece of evidence that A. afarensis was adapted to walking upright on the ground.

During the 1970s A. afarensis was the most primitive known species of early hominid, and we concluded that it occupied a basal position on the human family tree as the ancestor to all later hominids.

Broadly speaking, we put forth the hypothesis that all later species of hominids, including those of *Australopithecus* and *Homo*, could trace their origins back to *afarensis*. Picture a diagram shaped like a *Y*: the stem of the *Y* is *A. afarensis*, the ancestral hominid species, and the two limbs represent two distinct evolutionary branches, one leading to later species of *Australopithecus* and the other giving rise to *Homo*. Depicting evolutionary relationships in this manner is called a family tree, or more properly a phylogenetic tree. Such a tree supposedly reflects the ancestor-descendant relationships between different types of hominids. This phylogeny is the most widely embraced rendering of early hominid evolutionary history and thus identifies *A. afarensis* as *the* ancestral species to all later hominids. In essence this meant that *A. afarensis* occupied a pivotal place on the tree between more ancestral, more apelike hominids and more evolved—or, as anthropologists say, more derived—species.

Traditionally paleoanthropologists determined the relationships between fossil hominid species based on the overall similarities or differences between them. This often resulted in phylogenies that were rooted more in preconceived, intuitive notions of how the family tree should look and less in objective scientific methodology; therefore, it was almost impossible to evaluate the veracity of one versus another. Because of the proliferation of family trees into a forest, it was not possible to see the trees for the forest.

In the early 1950s Willi Hennig, a German scholar, proposed a sophisticated and more objective approach that he submitted was rooted in solid science, not intuition. Hennig believed that his method would generate a more objective understanding of the evolutionary history of species that could be evaluated and adjusted in a scientific manner. Unlike a phylogenetic tree, his approach ignored time and simply looked at specific traits or anatomical characters. He translated the occurrences of these features, some of which he considered ancestral and others as derived, into branching diagrams. He called his analytical method phylogenetic systematics. But Ernst Mayr, focusing on the branching nature of Hennig's diagrams, referred to them as cladograms, using the Greek *klados,* meaning "branch." The cladograms represented the proper ordering of ancestral and derived anatomical characters. Because a number of competing trees could be constructed, Hennig employed the notion of parsimony; that is to say, the tree that depicted an arrangement of the traits demanding the least number of evolutionary changes between primitive and derived was the simplest and therefore the one that most accurately represented the evolutionary relationship between the different species.

Late one afternoon Bill called me into his lab at IHO to view the results of his cladistic analysis. Each time I enter his lab I'm greeted by a black-and-white photograph of his intellectual hero, Franz Weidenreich: Smiling, bald, bespectacled, he sits at a lab table surrounded by skulls. A Jewish anatomist who fled Nazi Germany in 1935 and landed a job at the Peking Union Medical College, he dedicated his life to describing the *Homo erectus* fossils found at Zhoukoudian, eventually authoring one of the true bibliolandmarks in all of paleoanthropology—a monograph on the Peking Man fossils titled *The Skull of* Sinanthropus pekinensis, a copy of which perpetually resides on Bill's lab table as a reference and a source of inspiration and aspiration. (This, in fact, served as the model for our 254-page monograph on the A.L. 444 skull, not surprisingly titled *The Skull of* Australopithecus afarensis.)

"Whatcha got?" I asked.

Bill explained that he had selected eighty-two anatomical characters from the skull and dentition and run them through MacClade, a software program that processes and evaluates the phylogenetic relationships of physical characteristics. He had also used an additional software application that chooses the most parsimonious cladograms—those that require the least number of evolutionary steps between ancestral, primitive traits and the more derived ones. What I wasn't prepared for was the number of possible evolutionary trees Bill came up with—twenty-five. The good news was that when he examined the five most parsimonious ones, he found that *A. afarensis* was the sister species to all other species in these five cladograms.

"I can't say which is the best choice," Bill said. "But to me this demonstrates that *afarensis* is the ancestor to all later hominids."

Feeling giddy, I left his lab that afternoon with the assurance that our return to Hadar had not only resulted in the recovery of many new hominid specimens, including some outstanding skull material, but yielded important information about anatomy. And after thoroughly analyzing the data within a cladistic framework, we'd provided support for the theory that Lucy's species was pivotal in human evolution. Moreover, our strategically conceived approach to solving the outstanding problems facing us at Hadar after the work there in the 1970s made *A. afarensis* the single best understood early fossil hominid species.

CHAPTER 7

Lucy's World

Lucy lumbers along the lake margin at sunset, her bipedal gait accommodating elongated arms. When she stops and turns, I study her projecting face, small skull, and sloping forehead, and can see why people consider her a "missing link." She is precisely the transitional form you'd expect, a perfect blend of ape features and characteristics associated with modern humans. Crouching, she scoops a handful of fresh water, while I watch, mesmerized.

"Cut!" barks the director of the *NOVA* series *In Search of Human Origins,* and Lucy, who is really the actress Ailsa Berk in a rubber *A. afarensis* costume, stands and waves at me. Ailsa, who played Greystoke's ape mother, Kala, in the movie *Greystoke: The Legend of Tarzan, Lord of the Apes,* is perfect for this role, because she is a choreographer who specializes in primate movements. Yes, Hollywood has those.

It takes me a few seconds to return to reality, and I manage to nod when Ailsa says hopefully, "How did it look?"

I have replayed that scene in my head a hundred times, especially when I summit an Afar peak for the first time at the beginning of a Hadar field season and survey the landscape. The endless sand and punishing heat are so different from what Lucy experienced when she lived here more than 3 million years ago. Back then she probably spent nights in the protective canopy of the forest, her sleep occasionally disrupted by hippo grunts or the roar of a dinotherium, a peculiar-looking extinct elephant with tusks that pointed down to the ground instead of up. The air would have been cooler, a large lake would have shone in the distance, and she would have had access to a variety of habitats, from woodlands to patches of grasslands, where a great diversity of animals from squabbling prehistoric pigs to elegant giraffes kept a watchful eye out for ferocious saber-toothed cats and leopards. Yet even large felines were no match for the determined troops of hyenas who circled, growled, and eventually intimidated solitary hunters into retreat. Curled up and secure in a night nest, high up in a fig tree, Lucy probably had a front-row seat for this performance almost every night. Was she frightened? Did she comfort a baby in her arms? There were so many questions.

Of course, I am not the first paleoanthropologist to ponder the emotional capacity of human ancestors. Raymond Dart, the discoverer of the first *Australopithecus* fossil, promoted a picture of snarling killer apes that slaked their "ravenous thirst with the hot blood of victims and greedily [devoured] livid writhing flesh." Dart's grisly portrayal was perhaps in large part a reaction to the cruelty and viciousness of World War II, which showcased the evil side of humanity. But the animal bones Dart believed had been used in killing turned out to be leftovers from hyena meals and gnawed bones accumulated by porcupines. Bob Brain, the director emeritus of the Transvaal Museum in South Africa, also documented carnivore damage on the australopithecine bones themselves, particularly puncture wounds on skulls, presumably made by leopards who killed and dragged their prey into trees. In contrast to Dart's scenario, Brain concluded that the australopithecines were the hunted, not the hunters, a vision more in line with the interpretation proposed by Louis Leakey, the bighearted son of English missionaries, who saw our ancestors as gentle, loving, and kind, and living in peaceful cooperation with their fellow hominids.

Which view is correct? Fossils can tell us about anatomy and diet, but when it comes to emotion and behavior, we must turn to our closest relative, the chimpanzee, whose DNA differs from ours by only 1.2 percent. In fact, we are so closely related to these creatures that we have been variously called the "third chimpanzee," the "bipedal chimpanzee," and the "naked chimpanzee."

Described as tailless, long-armed, large-eared, humanlike beasts by Portuguese sailors in the 1600s, chimps had to wait almost a century before earning a more dignified identity when the English physician Edward Tyson came into possession of a carcass. "Wholly a Brute," he determined, "tho' in the formation of the Body, it may be more resembling a Man, than any other Animal; an intermediate Link between Ape and Man." More than 150 years later, British anatomist Thomas Henry Huxley concluded in his *Evidence as to Man's Place in Nature* that we and the African apes, especially the chimp, shared an ancestor that may have possessed certain behaviors common to chimps and hominids. In 1925 Robert Yerkes, a pioneer in the study of primatology, wrote, "Chimpanzees manifest intelligent behavior of the general kind familiar in human beings . . . a type of behavior which counts as specifically human." As early as the 1960s, chimps were seen using saliva-moistened sticks to coax delicious termites out of termite mounds; the first evidence of tool use in any primate besides man. And in 2007 Paco Bertolani, a graduate student at Cambridge University, and Jill Pruetz, an anthropologist at Iowa State University, reported having observed chimpanzees making and using wooden spears to hunt small primates called bush babies that are nocturnal and sleep during the day in holes in trees. Most often during the rainy season female chimps systematically search those hollows in trees for the presence of sleeping bush babies and employ twigs they have sharpened with their teeth as spears to stab and kill these small primates. Because the creation and manipulation of tools has long been considered the benchmark of hominid activity, these discoveries have blurred the line between humans and apes.

So how exactly does our closest living relative compare to our distant ancestors? Consider that, while *A. afarensis* flourished for close to a million years—a successful stint in anyone's book, and one that gave rise to even more effective, larger-brained bipeds—chimps in their 8 million or so years of evolution since their separation from our common ancestor have experienced far less dramatic changes.

Everyone knows the story about how an earnest twenty-three-year-old woman named Jane Goodall approached Louis Leakey after a lecture in 1957 and told him that she, too, believed that chimpanzees held many important clues about primitive hominid behavior. Impressed with her passion, Leakey took her on as an assistant, and eventually encouraged her to undertake fieldwork at Gombe National Park on the eastern shore of Lake Tanganyika to observe chimpanzees in the wild.

Jane's early field observations revealed the gentle side of chimpanzee society, where emphasis was placed on the kindness showed by chimps to one another. Consistent with the "Make Love, Not War" movement in the 1960s, chimp society was often held up as an example of a utopialike community populated by peace-loving vegetarians. Affable chimps were described as engaging in a wide array of peaceful behavior, such as hugging and kissing. Social bonds were reinforced with extended periods of grooming. Mothers were tender to their offspring, willingly adopting orphans. When aging or injured members of a tribe were dying, sympathetic cohorts would bring food or just linger nearby in a show of support or comfort. At Gombe National Park, Jane Goodall watched a chimp die of grief following the death of its mother. It was widely suggested that the idyllic, peaceful world of our closest living relatives was a testament to the level to which human society had sunk. We "civilized" animals were killing people in Vietnam. Why couldn't we be more like chimps?

That peaceful portrait of chimp society was fractured in the 1970s when researchers observed savage killing, infanticide, and even cannibalism in free-ranging chimps. Male chimps regularly form close alliances and patrol the periphery of their territory, resembling none so much as vicious gang members who hunt down and kill unfamiliar males and infants. When a female with offspring joins a new troop, the resident males systematically kill her young. This prevents the male infants from maturing and introducing novel genes into the group, and because a lactating female does not get pregnant, the elimination of her offspring brings her back into estrus, whereupon the resident males can impregnate her. Could such destructive and repulsive behavior so typical of human and chimp societies have been part of *A. afarensis* society?

Richard Wrangham of Harvard University, a leader in chimp research, examined the frequency of violent behavior resulting in death in humans and chimps and concluded that the rates were comparable, but that the level of nonlethal aggression—such as inflicting bite wounds—was significantly higher among chimps. It may very well have been the case that *A. afarensis* males fervently protected their territories with border patrols.

Scientists have established that female chimps leave their natal group to join other troops of chimps. This pattern is typical of modern human societies as well, and may have likewise characterized that of *A. afarensis*. Such behavior may impart a reproductive benefit because it avoids the downsides of inbreeding, which reduces the fertility of offspring. Males, in contrast, remain in their natal group and are therefore more closely related to one another than females are. This, along with their lifelong association, promotes very strong social bonds between male chimps that hang together and form strong alliances. Bonded males are thus encouraged to cooperatively defend their territory, a behavior that is fairly rare among mammals.

In 1999 prominent primatologists observing chimps at six different sites were asked to record behaviors that qualified as cultural traditions. They reported thirty-nine distinct behaviors, including no fewer than nineteen types of tool use. Teaching is one of the most important elements in establishing a culture, and chimps learn early that they can use twigs for capturing termites or ants and wad up leaves to use as sponges for soaking up water. They learn from their elders how to retrieve protein-rich bone marrow from carcasses and break open hard-shell nuts and fruit by using one rock as a hammer and another as an anvil. These are clearly not instinctive behaviors, but resourceful traditions passed down from generation to generation.

Furthermore, different groups of chimps have distinct cultures. Members of the Taï culture in Côte d'Ivoire, for instance, knuckle-knock on trees to attract females; they employ short sticks to eat ants and extract marrow from bone; and they crack open nuts with a stone-and-anvil technique. Chimps of the Kibale Forest in Uganda, meanwhile, give each other "high fives" upon greeting, but do not collect bone marrow or crack open available nuts. Yet some behaviors are far more widespread, such as tearing up leaf blades to attract playmates or fertile females and wildly dancing during a heavy downpour (rain dancing).

Although such patterned behavior falls short of human inventiveness and creativity, chimps do exhibit rudimentary culture. I was immeasurably impressed by a video clip posted on the Internet that showed a chimp investigating a termite mound. Usually termites escape from their home through an opening on the outer surface that leads through a small tunnel into the mound. But this particular termitarium lacked any openings. The chimp held a slender twig in its mouth and used a thicker stick as a probe to force into the termite hill so he could access the buried insects. When the stick met with resistance, the chimp used his foot to force the stick in deeper, the same way a human would manipulate a shovel. He did this several times, each time sniffing the end of the probe for evidence of termite odor. Eventually he located the bounty, at which point he put down the "digging stick" and used the more flexible twig he had been carrying in his mouth to withdraw his prey. The chimp showed considerable forethought to search for a meal, and he even brought the necessary "tool kit" along to complete the job. Whenever I show this video in class I can't help envisioning early hominids employing similar tactics.

Meat eating has played a central role in human origin studies, and our predatory behavior has often been invoked to explain early human cooperation, division of labor, brain expansion, and increased intelligence. Jane Goodall was among the first to document preda- tion and meat eating by Gombe chimps—behavior that came as a minor revelation to many anthropologists, who were surprised that a vegetarian ape would actively pursue red colobus monkeys, bush pigs, and small antelopes. The initial response was that this was aber- rant behavior provoked by Jane's providing bananas for wild chimps. But as more and more observations of chimps hunting were reported, it was inescapable that systematic hunting of small and medium mammals is typical of chimps. Even more astonishing is the fact that the amount of meat consumed in some chimp communities is not that much lower than is typical in some living hunter-gatherer societies. The pattern of hunting behavior differs among chimp communities. Gombe chimps hunt independently, but Taï chimps, which have been extensively studied by the Swiss primatologist Christophe Boesch of the Max Planck Institute for Evolutionary Anthropology in Leipzig, Germany, hunt in groups, with some males strategically driving the prey toward other males while others block potential escape routes. Another surprising feature of the process is that male chimps share meat, something that is almost unheard of outside human societies. It's possible that meat is shared between males who have partaken in a cooperative hunt in order to reinforce male-male bonds, ensuring alliances that might be useful during future hunts. Hunting occurs even when fruit is available so it is not driven by hunger.

Chimps seldom scavenge, and driving off a predator such as a saber-toothed cat would have been highly dangerous for *A. afarensis*. Lacking agility and speed, as well as large, slashing canines, *A. afarensis* would also have been ill equipped to chase and capture prey. Some hunter-gatherers can exhaust an animal by continually pursuing it, but the high caloric cost to *afarensis* and the increased possibility of injury or even death would exceed the rewards.

While chimp hunting is based on pursuit and capture of smaller prey with their powerful hands, human hunting tends to focus on a larger prize, which necessitates the use of some sort of weapon. But this distinction between hunting techniques faded when Bertolani and Pruetz filmed Senegalese chimps using mini spears made of sharpened sticks to kill bush babies in their daytime sleeping nests. Certainly *A. afarensis* would have been capable of similar behaviors to procure small, slow, or sleeping game. However, meat would not become a significant proportion of the hominid diet until the invention of stone tools. Without slashing teeth or stone implements, *A. afarensis* would not have been capable of capitalizing on the meat of a large mammal, even if they succeeded in driving off carnivores from a kill.

Muscular and athletic, *A. afarensis* would have been able to climb trees not only to avoid predators but also to search for fruits, seeds, birds' eggs, and other foods. Rising at dawn, *afarensis* probably foraged in the surrounding landscape, perhaps even beginning with a morning snack of fruits or nuts in the very trees in which they slept. With few exceptions, primates, and indeed most mammals, live in social groups. I suspect that Lucy—like monkeys, apes, and humans—engaged in fairly complex social behavior. Group size varies considerably among primates, but modern-day hunter-gatherers congregate in groups of 150 or so individuals, not much larger than troops of chimpanzees. I suspect that from time to time, like living chimps, *A. afarensis* groups would break up into smaller foraging subgroups so as to reduce feeding competition and aggression, then regroup at the end of day. This behavior, often referred to as "fission-fusion," is typical of chimps and may have been part of Lucy's social world. Whereas females may have formed larger foraging groups for safety in numbers, perhaps *A. afarensis* males, not unlike chimps today, preferred to search for food in all-male bands.

It is impossible to know the precise diet of *A. afarensis,* but chances are they subsisted on a broad range of plant foods, such as nuts, fruits, and maybe even roots and tubers. As opportunistic feeders *A. afarensis* might also have sought out honey to satisfy a sweet tooth, just as chimps do today, and snacked on birds' eggs or even dug up turtle and crocodile eggs like the ones preserved in strata near the Lucy site. The relatively large incisors, especially in males, indicate a fruit-dominated diet, much like a chimp's, but the lack of enormous crushing and grinding back teeth argues against consumption of large amounts of tough, fibrous vegetation. Still, the enamel cap covering *A. afarensis* teeth is relatively thick—an obvious adaptation for extending the life of their teeth. In *A. afarensis* the canine teeth have undergone considerable reduction, especially when compared with those of chimps and gorillas, and would not have played a major role in ripping open tough vegetation or slicing through meat. But microscopic observation of the lip side of some canines shows classic chippage, indicative of biting hard objects like nuts.

Much of my musing about the behavior of *A. afarensis* is based on the large body of behavioral data that has been gathered for chimpanzees in Africa. Perhaps many of the behaviors characteristic of those apes are ones that were inherited from the common ancestor. To address that possibility, Richard Wrangham decided to tabulate which behaviors are seen in humans, chimps, bonobos, and gorillas. He surmised that if a behavior is common to all four species it was probably present in the common ancestor and presumably also in early hominids such as *A. afarensis.* His results indicated to him that early hominids probably lived in closed societies dominated by aggressive males that guarded their territory from outsiders, with females emigrating from the community in which they

were born. This does not mean, however, that *afarensis* did not devi-
ate from this suite of ancestral behaviors. *Afarensis* must have possessed their own unique set of behaviors because the species was a
provocative blend of humanlike features, such as bipedalism, and
apelike characteristics, such as a small brain, short legs, and so on,
that must have had an unknowable impact on how they procured
and processed food, how far they may have wandered each day, and
how they interacted with one another socially.

Social behavior simply does not fossilize. If we found fossil
remains of the other chimpanzee in Africa, the bonobo, we would
have no way of knowing about the unique behavior of this species
from the bones alone. More lightly built than the other chimps, the
bonobo, confined to the Democratic Republic of the Congo, has
been known since the late 1920s. The American anatomist Harold
Coolidge designated it a distinct species, *Pan paniscus*, in 1933, and
observed that it "may approach more closely to the common ancestor of chimpanzees and man than does any living chimpanzee."

The slender bonobo has elongated legs and has been observed
to stand and walk bipedally more frequently than the chimpanzee.
Like chimps, bonobos live in a fission-fusion society. But among
bonobos, it is the females who show strong bonding, not the males.
While there is some hostility between bonobo groups, they more
often peacefully intermingle than do chimpanzees, for whom intergroup avoidance and aggression is the rule. In contrast to chimpanzees, the level of violence among bonobos is fairly low, and
cannibalism, infanticide, and intergroup warfare, have not yet been
observed. Although not entirely peaceable, bonobos stand out
because of the high frequency of reconciliation between animals following aggressive interactions. Like chimps, bonobos kiss and caress
during reconciliation, but remarkably they also engage in a wide
range of sexual behaviors. Males will mount one another, rub their
scrotums together, and even fence with their penises. Females participate in oral sex, tongue kissing, and mount one another face-to-
face to rub their genitals together. Sexual activity among bonobos is

unlike that of any other nonhuman primate, particularly during feeding times, which are accompanied by nothing short of an orgy.

In contrast to human females, who are sexually receptive throughout their sexual cycle, chimpanzees restrict intercourse to period of estrus. Female bonobos, however, are sexually active during much of their cycle and often employ sex for certain favors, like obtaining food from a male. Bonobos are even more humanlike in their frequent choice of a face-to-face copulation position, perhaps reflecting the more forward positioning of the clitoris and frequent female-female rubbing.

Female bonobos form strong alliances and are the ones in control of bonobo society, driving off males when they become aggressive and threaten other animals. Other apes use their large body size or threatening canines during such aggressive encounters. But female bonobos have small bodies and small canines. This and many other aspects of bonobo society attest to the variability of ape behavior and highlight behaviors that would not be discerned from the bony anatomy of these apes, thus complicating any attempt to reconstruct sexual behavior in *A. afarensis*.

What, if anything, can be said about mating behavior in *afarensis*? Generally speaking, primate species that show little difference in canine or body size between the sexes tend to be monogamous. Size differences between males and females—sexual dimorphism—usually mean that these animals live either in harem groups like gorillas or in more sexually promiscuous groups like chimps and bonobos. With this in mind we can look at *A. afarensis*, which is highly sexually dimorphic in body size, with males being much larger than females. The ratio of male to female body weight provides an estimate of the level of dimorphism in a species. Humans have a ratio of 1.22; chimps and bonobos come in at 1.3; and in gorillas the ratio is over 2, reflecting the fact that gorilla males are twice the size of females. The estimated ratio for *A. afarensis* is noticeably high, and at 1.5 exceeds that of chimps and bonobos.

The higher level of body-size sexual dimorphism in *A. afarensis* may be part and parcel of living on the ground, because terrestrial primates have larger male-female size differences. Males are generally responsible for group protection and size is an advantage for driving off danger. The smaller size of females is probably favored because of the critical energy demands placed on pregnant and lactating females. With generally low-quality vegetarian-dominated diets it would be a strain for females of large body size to consume sufficient food to meet those reproductive demands. *A. afarensis* is peculiar among primates because body-size sexual dimorphism is high, but canine dimorphism is low, with both males and females possessing small canines. Apes use their large canines during aggressive displays and often inflict deep and sometimes lethal wounds on one another. *A. afarensis* thus presents something of a quandary when anthropologists attempt to reconstruct mating behavior. Reduced-canine sexual dimorphism suggests less competition for access to estrus females, but the large body size differences suggest the opposite.

Darwin explained that the reduction in human canines in both males and females was a consequence of the invention of stone tools that acted as a replacement for canines as weaponry. It's a fascinating hypothesis, given the fossil information available at that time. Subsequent discoveries have proven him wrong, however. Australopithecines, including *A. afarensis,* with its reduced canines, did not make stone tools but may have used weapons made of perishable material such as wood.

Perhaps the reduction in hominid canine size is due to a decrease in male-male mating competition. This idea is favored by Owen Lovejoy, who has even postulated that in *A. afarensis* we see the beginnings of monogamy, an idea that has come under considerable debate, especially because there are a number of modern human societies in which monogamy is not the accepted norm. Charlie Lockwood investigated the relationship between sexual dimorphism and mating behavior in *Australopithecus robustus,* a species known only from South Africa. He determined that although canine sexual dimorphism was absent, with both males and females possessing small canines, body-size sexual dimorphism was considerable, with males weighing two times more than females, not unlike the level of sexual dimorphism in *A. afarensis*.

The bulk of Charlie's sample was from the Swartkrans cave, where work by Bob Brain concluded that leopard predation was the major factor responsible for the accumulation of the hominid fossils. The Swartkrans sample consists predominantly of young adult males, and Charlie believed that they suffered high predation rates because they were living on the periphery of the main social group consisting of a silverback *A. robustus* and his harem. Younger males could not compete with the large males until they fully matured—something that apparently occurred later in males than females. Primates such as gorillas that exhibit this pattern of bimaturation, in which males have extended growth periods, are associated with intense male-male competition for reproduction.

Could this mean that *afarensis* lived in a polygynous mating structure? For the moment I am not committing to a specific mating structure for *A. afarensis*, but based on the remains at the First Family site, it appears that an *A. afarensis* group consisted of adult males and females, ruling out a harem group. I am inclined to think that *A. afarensis* lived in multimale-multifemale communities where females may have mated with more than one male, contrary to Lovejoy's idea of monogamy. Monogamy would be a strange choice considering that, humans aside, no primate that lives in a larger social community embraces monogamy as a mating strategy.

Today Hadar is a desert, but we know from the animal fossils that have been collected over the years from the Hadar Formation that the habitat in which Lucy lived was dramatically different. These nearly eight thousand fossils, ranging in size from rodents to elephants, line the shelves at the National Museum of Ethiopia, and although they don't share the same star billing as Lucy, they offer us a vital glimpse into the habitat and paleoenvironment of *A. afarensis*. Experts from around the world have made pilgrimages to the museum to study particular types of animals, pigs, elephants, rhinos, antelopes, giraffes, carnivores, monkeys, and so on. Kaye Reed, the Hadar Research Project's paleoecologist, looked at the collection as a whole. A confident redhead with an explosive laugh and a wild streak, Kaye is great company in the field. Her goal was to tease out an understanding of the habitats in which *A. afarensis* lived and also to see if there were any major changes in ecology within the Hadar Formation over time.

Kaye is especially well qualified for this task because of her wide-ranging African travels, where she acquired firsthand knowledge of the preferred habitat for many living African species. Kaye thoroughly compared the fossil assemblages from Hadar with those found in a variety of African habitats today, using the principle of uniformitarianism, which states that if the types of animals are similar the habitats must also have been alike. Paleoanthropologists have long been interested in reconstructing the paleoenvironments in which fossil hominids lived to gain insight into paleobehavior, their extinction, and even speciation events that gave rise to new species.

In her analysis, which was published recently in the *Journal of Human Evolution*, she compared the composition of the fossil collections from different levels within the Hadar Formation to modern African habitats populated by similar animal species. She considered twenty-three modern African settings found in national parks and reserves that ranged from forests to deserts. The beauty of Kaye's approach is that she equated fossil habitats at Hadar with exact modern African analogues that one can actually visit on safari and thereby experience the landscape that would have been familiar to *afarensis* millions of years ago. If you have been on an African safari, or watched lots of nature programs on TV, you know that the kinds of animals that are typical of the savanna grasslands like the Serengeti Plain are very different from those found in wooded habitats like the eastern Transvaal of South Africa. Whereas the Serengeti grasslands are dominated by vast herds of wildebeest, zebra, gazelles, prides of lions, and cheetahs, wooded areas like those in the Transvaal are populated by elephants, kudus, leopards, monkeys, and giraffes.

By comparing the fossil vertebrate species collected from the older geological levels at Hadar, Kaye concluded that early *A. afarensis* lived in a mixed woodland/grassland habitat not unlike the Okavango Delta in northwestern Botswana, where wide grassy floodplains and swampy areas are bordered by forests beyond which lie more open savanna grasslands. In contrast, the faunal assemblage recovered from the later level where Lucy had been buried more closely approximated Kenya's Amboseli National Park, where open grasslands dominate in a more sparsely treed and much drier landscape.

Kaye looked at twelve geological horizons within the Hadar Formation and determined that they sampled a variety of habitats, including grasslands, woodlands, and even dense woodlands; *A. afarensis* fossils turned up in all of these habitats. This diversity of environments suggested that they and many of the other vertebrate species Kaye identified were very adaptable creatures roaming and surviving in a wide range of environments. As a generalization Kaye characterized the paleoenvironment throughout the Hadar Formation as a fluctuating woodland habitat, with some open patches. Paleoenvironmental reconstruction of Laetoli in Tanzania, another *A. afarensis* site dated to 3.6 million years, suggests that the habitat there was less well watered but heavily forested. From the diversity of habitats in which *A. afarensis* lived it appears that this species was highly flexible, at ease in many habitats that ranged from closed woodlands to more open shrublands.

To a very large extent understanding the geology of Hadar was invaluable for creating a total picture of what Hadar had been like when *A. afarensis* flourished there. Chris Campisano, a young burly scholar, joined the Hadar Research Project in 2001 as a geologist. Now a postdoctoral candidate at IHO, he was a former student of Craig Feibel's at Rutgers University. When Craig and Chris first came to the field it was astonishing to see their facility for reading Hadar's geological record—like watching Egyptologists translating hieroglyphics.

Chris took to the outcrops with a zeal and an eagerness that was impressive, and over the course of just a few field seasons he synthesized the stratigraphy, argon dating, and history of deposition of the Hadar Formation into a framework that has become a comprehensive and authoritative statement on the Hadar Formation. He has delivered just the geological road map we need to fully comprehend Lucy's world.

Probably due to the strong influence of his mentor, Chris, like many paleoanthropologists, was fascinated by the widespread discussion about the potential influence on human evolution of global climatic change. From an evolutionary perspective changes in climate generate changes in habitat, and creatures under these new conditions, including hominids, have three choices: adapt, perhaps even evolve into a new species; migrate into areas that have not changed; or go extinct. Chris was intrigued by the notion that *A. afarensis* had not changed dramatically over the course of its 500,000-year existence, so he and Kaye began to work together to investigate whether there was any evidence at Hadar for geological and environmental change that might have resulted in significant evolutionary events among hominids. Chris's intimate knowledge of the Hadar strata suggested to him that the geological record at Hadar might contain clues to questions of the connection between climate change and hominid evolution. The precise chronology of the sediments at Hadar, the abundant and diverse collection of fossil hominids and other vertebrates, as well as the geological evidence proffered a unique background against which to view that plausible connection.

Chris was especially interested in a geological feature that he saw near the top of the Hadar Formation called a disconformity. A disconformity is a surface that separates younger and older strata and represents a time of nondeposition, a convenient horizon that has been used to assign the sediments above to the Busidima Formation, named after a local riverbed. He observed that the Hadar Formation sediments beneath the disconformity were fine-grained sands, clays, and silts that were deposited, relatively slowly, in rivers and lakes. The Busidima sediments just above the disconformity were very different and included collections of pebbles and gravel cemented together in a rock called a conglomerate that were deposited swiftly, presumably by the paleo–Awash River. Roughly 200,000 years are missing—a vital slice of time, as we will see, between 2.7 and 2.9 million years ago.

Chris and Kaye then began to look closely at the kinds of animal fossils there, especially antelopes because they are abundant in the fossil record and habitat specific. Throughout much of the Hadar Formation the antelopes indicate a mosaic set of environments, as Kaye had concluded. But those from the Busidima Formation suggested a major shift in habitat to dry open environments dominated by grasslands. The antelopes from the lower Busidima Formation were grass eaters, such as hartebeests, wildebeests, and gazelles that were at home in a tall grassland habitat. In the upper reaches of the Hadar Formation Chris and Kaye noted the presence of some open-habitat antelopes that suggested to them that the environmental change was already under way prior to the level of the disconformity.

Chemical analyses of some of the Hadar fossils bolster the hypothesis that there was a shift in habitat. All plants take in carbon from the environment during photosynthesis. Some classes of plants, such as tropical grasses, utilize a specific photosynthetic pathway that favors accumulation of the carbon-13 isotope in a process called C4. Other types of plants, such as shrubs and trees, use a different photosynthetic process called C3 that does not favor carbon-13, resulting in lower concentrations of this isotope. By analyzing only a few milligrams of a fossil tooth or bone we can determine whether that animal ate C3 or C4 plants. For example, fossil horse teeth have enriched levels of ^{13}C consistent with the grazing lifestyle and C4-photosynthetic pathway typical of living horses. In contrast, the bones and teeth of extant and fossil kudus bear a C3 signature, in keeping with their tendency to eat trees and shrubs.

One of our Ethiopian Ph.D. students, Million HaileMichael, has performed preliminary carbon isotope analysis on animal fossils from the Hadar Formation. Million's study of carbon levels in the paleosols were consistent with Kaye and Chris's faunal and geological studies. The low carbon-13 levels below the disconformity reflected a landscape made up of 70 percent trees and 30 percent grasses. Above the disconformity, however, the remains have high carbon-13 levels, indicative of habitats containing at least 50 percent grasslands.

Paleoclimatologists have long noted a global cooling and drying trend throughout this time range, and Hadar seems generally to substantiate this. The real question is just how that global climate change may have influenced vertebrate, and especially hominid, evolution. The antelopes that Kaye and Chris concentrated on reflect a major vegetation change in which drier-adapted species were favored. But what about the hominids?

That is what is so incredibly interesting and provocative. Below the disconformity in the Hadar Formation we find only *A. afarensis* and no stone tools, whereas above the disconformity *A. afarensis* is nowhere to be seen, and stone tools make an appearance along with the oldest known evidence of *Homo*, an upper jaw, dated to 2.4 million years. Something big had happened during the missing 200,000 years: The paleoenvironment changed dramatically, *A. afarensis* vanished, and a new kind of hominid, *Homo*, took its place. What happened during those missing years that might explain the demise of *A. afarensis* and the appearance of *Homo* is a mystery, but it sure looks like it might be related to the change from a wetter climate to a drier one.

Growing Up Australopithecine

Frequently I am asked if Lucy had given birth. It's a logical question, because she was female and the erupted and partially worn wisdom teeth indicate that she was an adult when she died. Any woman who has given birth knows that it's a painful and traumatic experience. During parturition the pelvis is subjected to extraordinary forces as the newborn passes through the birth canal. Quite often the most forward part of the pelvis, the pubic area where both halves of the pelvis meet, separates slightly, damaging the bone surfaces and leaving pits or scars in the bone. When a forensic pathologist examines a female skeleton and sees these "scars of parturition" the assumption is that the woman had given birth at one time or another. Lucy's pelvis shows no signs of damage in her pubic bones. Does that mean she hadn't been a mother? Not necessarily.

We really cannot tell from her bones, but I imagine that an apparently healthy, young female *A. afarensis* like Lucy would in all likelihood have reproduced. When we look at the demographics of a chimpanzee troop a very high percentage of females become pregnant soon after they reach sexual maturity. So it is a pretty good guess that Lucy did have a kid or two. Recently, a stunning find from a site across the river from Hadar has revealed what Lucy's child would have looked like, if she'd had one, and given us an unprecedented glimpse of childhood in an early human ancestor.

The story began in late 2000, when a young Ethiopian scholar named Zeresenay "Zeray" Alemseged arrived at Arizona State University, where I had moved the Institute of Human Origins in 1997. He had come to take up a post as a postdoctoral research associate at the institute. A tall, handsome man, with a quiet resolve and a palpable pride in his homeland, Zeray was clearly a well-trained and rigorous researcher, and I knew he would go far. But I didn't anticipate how swiftly he would make his mark on the field. For quite some time, Zeray had had his sights set on working in an area adjacent to Hadar known as Dikika. With funding and field equipment from IHO, he was able to set off to Dikika that year with a small expeditionary group. He was determined to find hominids, and when he left IHO he told me, "You won't be disappointed."

And indeed I wasn't. I learned in an e-mail that Zeray's team had made a stunning find on Sunday afternoon, December 10, 2000. Tilahun Gebreselassie, the antiquities officer assigned to Zeray's expedition, spotted a cheekbone eroding out of a sandstone horizon. Zeray knew immediately that it was the cheek of a baby hominid skull because of the small canines, lack of brow ridges, and the vertical chin. He quickly scanned the area and found a forehead bone fragment. Having only three workers and no excavation equipment, he decided to return to Addis Ababa. He knew it was critical that such a precious find be placed in safekeeping at the National Museum.

After returning to DIK-1-1, as the locality was designated, Zeray and his small team made a thorough search of the hillside for additional fragments that might have already eroded out onto the surface. He collected and mapped all of the fragments and passed as much loose sediment as time would allow through a fine mesh screen. A more comprehensive excavation of the DIK-1 hillside would take place during the next four field seasons.

After Zeray returned to Addis Ababa he carefully examined his prize and arranged to have Alemu Admassu, a museum technician, make a mold of the baby skull so a record cast could be readied and safeguarded in the paleoanthropology laboratory. This first cast serves as a permanent record of what the specimen looked like when it was found. Fortunately for me, Zeray ordered a second cast of the specimen that he could bring back to IHO.

The day Zeray invited me to his little windowless office, just off the main research lab at IHO, he was sporting a big smile and holding a box containing a cast of his find. I sat down as he carefully unwrapped the soft, pink toilet paper that protected the cargo he carted back in his hand luggage from Ethiopia. Cast in white plaster, the shape of the skull was immediately obvious to me. Peering into the eye sockets, I knew I was looking at an *A. afarensis* baby. "It's Lucy's baby!" I exclaimed.

Zeray was guarded about calling it *afarensis,* and as he was explaining the extraordinary preservation and completeness of the find he said, "I'd rather wait until I have cleared away the matrix before I assign it to a species."

I thought being cautious was good. I was only thirty-one when I found Lucy, and as I look back I know there were some claims I made about the Hadar fossils that I had to later retract. For example, in my first *Nature* publication I suggested there might be as many as three different species represented in the Hadar sample. We now know that with the exception of one upper jaw assigned to *Homo,* all other hominid specimens at Hadar belong to *A. afarensis.* Every scientist must accept that mistakes are sometimes made and have to be corrected; that's part of the learning curve for all of us.

Zeray's tour of the plaster block was mesmerizing. I could see that the lower jaw, the mandible, was still in occlusion with the cranium. Zeray ran his fingers over the cast with great familiarity of the topography and pointed out what appeared to be vertebral spines and even some ribs. It was an exceptional moment for me when I realized Zeray had launched his career with a home run.

Zeray gently placed the cast in my hands and said, "Have a closer look, Ato Don." He always called me Ato Don, that was his nickname for me: Mr. Don in Amharic. As I turned the cast over and over in my hands I became more and more convinced that I was looking at a baby *A. afarensis.* Slightly dazed, I turned to Zeray and suggested that we go into the lab and compare it with the Hadar casts.

The main IHO lab is lined on two walls with tall, white wooden cabinets that contain our comparative collection of fossil hominid casts from around the world. We walked directly over to the cabinet marked Hadar and opened it. I surveyed the drawers and pulled out the one labeled "cranial remains." I was specifically searching for A.L. 333-105, a partial juvenile cranium collected at the First Family site in 1975. Although it was not as complete as the Dikika specimen, there was enough overlap in the preserved parts to make a meaningful comparison between the two.

Both individuals died at roughly the same developmental stage, when the two milk molars had erupted and the first permanent molars were still developing in their crypts. I pointed out a number of similar features in facial anatomy and vault shape and made an argument for the near identity of the two specimens. "There it is, I'm convinced we are dealing with the same species, *afarensis.* What do you think, Zeray?"

He had made this comparison earlier in Addis Ababa where the original A.L. 333-105 specimen resides in a safe, so little of what I said was new to him. Zeray looked at me and replied with his customary cautiousness, "I want to wait to see, Ato Don."

The cast convinced me that Zeray's find belonged to *A. afarensis,* but I was eager to confirm that by examining the original specimen back in Addis Ababa. Casts are very good representations of the originals, but I knew from experience that there is no substitute for working with the real thing. In June 2001 I left for Ethiopia to join my colleagues Bill Kimbel and Yoel Rak at the National Museum of Ethiopia for lab work on the Hadar hominids and, most important, to meet Zeray and see the original Dikika fossil.

When I walked into the small lab where Zeray was hunched over a binocular microscope, I could see the brownish gray specimen, no larger than a cantaloupe, under the bright light from the fiber-optic lamps above. Zeray turned off the lamps and gently placed the fossil on a yellow foam pad on the table in front of me. I sat down and asked him if I could touch the specimen. Being very careful to keep the fossil over the pad I picked it up and was delighted to see how pristine it was. The bone had a lovely patina and the dark gray surfaces of the teeth sparkled. It was a gorgeous specimen, far surpassing the plaster cast I had seen in the lab back in Tempe. As Zeray sat close by and outside light flooded in through the windows that lined the small room, I thought, The Dikika baby is the find of a lifetime for Zeray, just as Lucy had been for me.

Zeray had a mountain of work to do on the fossil before he would be able to see all of the anatomy. As it was, he had already spent quite a few hours photographing, studying, and preparing the specimen. The block of sandstone that encased the fossil was rock hard and contained an untold number of bones, the exact position of which was totally unpredictable. There was only one way of removing the sandstone matrix from the actual fossil bone and that was grain by grain. Here was a fossil so valuable that only a steady hand, long concentration, and enormous dedication could liberate it from the rock that surrounded it 3.3 million years ago. I knew Zeray was up to the challenge, but I didn't envy the amount of time he would have to spend in the lab squinting through a microscope.

After Zeray completed all of the time-consuming, but crucial, prepreparation of the specimen, he then developed a strategy for beginning the serious work of removing the matrix. The tool of choice consists of a small scribe, or pen, that holds a sharpened carbide tip, much like a drill bit. A hose connects the pen to a compressor and a knurled ring is rotated to regulate the flow of compressed air that causes the carbide tip to vibrate. I sometimes call this airscribe a microjackhammer.

Working in another lab at the paleoanthropology laboratory in the National Museum, I could hear the groaning compressor in the hallway and the constant buzzing of the airscribe coming from the room where Zeray worked. I envisioned him crouched over a microscope cradling the baby in one hand and the scribe in the other. By greatly restricting the airflow through the airscribe and using a relatively high magnification, one can remove individual grains of sand. Under the microscope the highly magnified view resembles the eroded landscape where the fossil was found. Carefully navigating the steep cliffs and valleys can be treacherous, and care must be taken to prevent chipping away too large a piece of matrix that might include part of the actual bone.

When a fragment of bone is inadvertently chipped away, one has to stop, turn off the airscribe altogether, locate the tiny flake of bone that might be only a few millimeters or less in size, then take a deep breath. A pair of microtweezers is used to pick up the chip: It has to be repositioned on the fossil and stuck on with a tiny drop of adhesive that is usually applied with the tip of a needle. No matter how careful one is, no matter how vigilant, these accidents happen, and the preparation procedure has to cease until the damage is repaired.

Zeray usually arrived early to the museum, sometimes ahead of me. I could hear him working in the lab across the hall and after a couple of hours he would emerge bleary-eyed from his lab table for a break. I knew from experience that the vibration of the scribe caused a tingling feeling in the hands while the tremendous concentration required while squinting through a microscope led quickly to fatigue. Taking a breather to relax and recharge is obligatory if one is to prevent any mistakes that might damage the actual fossil. A walk around the museum compound is one way to clear the mind, unless there is a torrential rainy season downpour. Taking a coffee break isn't a good idea because the caffeine causes one's hands to tremble, a movement that is magnified many times under the microscope, making controlled preparation a challenge.

I am sure only Zeray knows how many hours he has expended at the microscope cradling the skull, but he never complains. After all, it is his baby. He has to maximize his time in the Addis lab because the original fossils cannot travel abroad. The antiquities regulations forbid moving national treasures out of the country except under very special circumstances. Obviously this is to safeguard the original and reduce any opportunity for damage or loss. When I began work in Ethiopia in the 1970s, there was no paleoanthropology laboratory for working on fossil specimens, so we had to take them abroad. I felt the burden of responsibility when the early Hadar finds were under my care at the Cleveland Museum of Natural History and much prefer traveling back to Addis Ababa to work on the fossils.

During my yearly visits to Addis Ababa, as I continued my work on the Hadar hominids, I was able to monitor Zeray's progress as he slowly, painstakingly chipped away at the Dikika baby, revealing more and more of the specimen. One day in 2002 all of us working at the paleoanthropology laboratory were forced to take a break because of a power outage. Standing on the little veranda of the laboratory building, listening to the rolling thunder and watching the torrential downpour, I peppered Zeray with questions about the location and geology of the place where the baby had been found. He surprised me by asking, "Ato Don, why don't you come with me and visit the site yourself?"

With a big grin, I replied, "That would be terrific, Zeray."

In late November 2003 I joined Zeray's Dikika Research Project in the field for two weeks and saw the exact place where he had discovered the Dikika baby. The drive into the Afar was familiar to me, and I was pleased to see that the tarmac road, now called the Assab Road, was in good repair, making our trip faster than usual. We pulled into Zeray's camp around five in the afternoon, hot and sweaty after the ten-hour drive. After a lovely dinner of cannelloni, I crawled into my tent and slept until five thirty, when I was awakened by the sound of prayer coming from the local Issa people, enemies of the Afar I had so long worked with at Hadar.

Following a breakfast of French toast and coffee, I climbed into Zeray's Land Cruiser and we headed out to the baby locality. On the drive we passed a couple of large herds of camels being looked after by very young Issa boys who ran excitedly after our car. My anticipation grew as we continued our drive across a gravelly plateau toward DIK-1. We parked on the edge of the plateau and the Issa cautiously scanned the horizon for the presence of Afar tribesmen. With the all clear we descended into a canyon in single file, carrying the excavation equipment. Finally, after snaking our way along ridgelines and gullies for thirty minutes, we reached the place where the baby skull had been found. Everyone seemed to know his job, and Zeray's team began setting up screens and hauling buckets of loose soil down from the hillside.

After Zeray gave me a short tour of the locality and took me to the exact spot where the skull was found, he turned me over to his geologist, Jonathan Wynn of the University of South Florida. Soft-spoken and extremely bright, he has a remarkable knowledge of the stratigraphy. He took me to the top of the hill to show me where we were in the Hadar Formation. From that vantage point we had a good view of the surrounding area, and to the north I could clearly see Hadar. Jonathan took out an aerial photograph and pointed to our exact location, explaining that we were in the lower part of the Sidi Hakoma Member, just about 35 meters above the Sidi Hakoma Tuff and less than 50 meters below the Triple Tuff—the volcanic ash horizons we had argon-dated at Hadar. This meant the baby bracketed between 3.22 and 3.4 million years, making it roughly 3.32 million years in age, about 120,000 years older than Lucy.

As we walked back to the baby locality I could see swirling clouds of dust rising in the air and hear the gravelly sound of the sediments as the screeners slid the soil back and forth. Suddenly, one of the Issa workers held up a bone fragment and called out

Zeray's name. Zeray went over and examined the fragment. *"Truno, betam truno!"* he exclaimed. Good, very good. He placed the fragment in a tiny plastic bag and labeled it.

"What do you think it is, Zeray?" I asked.

"I think it is part of the baby, probably a fragment of its rib, which is great," he replied.

I sat down on the hillside under the glaring sun and watched the buckets of soil being carried down to the screeners. What most impressed me when I worked with Zeray was the strategy he had developed. After meticulously mapping the immediate area he collected and examined every spoonful of loose earth at the baby site. I am certain not the smallest bone splinter or enamel chip escaped the team's efforts.

This, I thought, is how the clues are found, the ones we use to reconstruct the story of our origins. Day in, day out, buckets of loose, dry soil are poured into a screen and bit by bit we collect a skull, a jaw, or a skeleton. Somewhere in Africa's Great Rift Valley, perhaps at this very moment, on some godforsaken landscape, a group of excavators huddled around a screen are waiting expectantly for the next bucket of soil and hoping, just hoping something important is found.

I was awakened from my reverie by Zeray's voice. *"Baka!"* Stop.

Everyone prepared to leave the excavation, stacking the buckets one inside another, collecting the trowels, hefting the shovels, and starting the long walk up and out of the canyon under the noonday sun. Huffing and puffing, kicking up the dust along the path, we finally reached the Land Cruiser and unlocked the door, letting the stifling heat escape before we settled into our seats and headed back to camp, where lunch was waiting.

The Dikika region had long held a strong allure for me. Its geological exposures were easily visible from Hadar and I knew they were merely a southern extension of the Hadar Formation, cut through by the Awash River. In 1974 I waded across the river with

Maurice Taieb and a number of other IARE colleagues to have a quick look at those geological exposures, and we recovered a partial hominid mandible cataloged as A.L. 277-1. This whetted my appetite for more surveying in the area, but when we tried to gain permission to work there, the Ministry of Culture told us that until we completed our work at Hadar and gave up our rights to that site, this would not be possible.

There was another very compelling reason why we could not work in the Dikika area. Hadar is Afar territory and Dikika is Issa territory. The rivalry between these two tribes has gone on for centuries, and I knew that we would never be able to bring our Afar colleagues to work in an Issa area. Even today both tribes regularly cross the river to raid each other's villages and shoot as many men as possible. With so much to do at Hadar, and out of loyalty to the local Afar, we did not give serious thought to working in Dikika. That was now Zeray's job.

One morning when I decided to take a break from the daunting work of excavation and screening at the site where the baby was found, I climbed one of the higher hills in the region. I could now gaze northward and soak in the brilliant expanse of Hadar. When I looked at the aerial photos I had in my knapsack, it dawned on me that the site where the baby was found was situated only 4 kilometers south of where I had discovered Lucy. Even more incredible was that I must have walked within a kilometer of DIK-1 when I visited this area in 1974. Was the baby out on the surface back then, just waiting for someone to spot it? I will never know.

On the drive back to Addis Ababa I reflected on my experience at Dikika, so generously offered to me by Zeray. Not only did I participate in the excavation and the recovery of more fragments of the baby skeleton, but I was also able to watch Zeray in action. His self-confidence and leadership skills make him highly respected and appreciated by everyone.

Unfortunately, at about this time, Zeray's postdoctoral position at IHO had come to an end, and he made the decision to accept a

position at the highly respected Max Planck Institute for Evolutionary Anthropology. We were sad to lose him, but it was time for Zeray
to move on. The Max Planck Institute offered him a research home
where he could concentrate his efforts on the analysis of the baby
skeleton and prepare the long-anticipated publication of the results
of his work at Dikika.

At long last, five years, nine months, and eleven days after it was first
found, the Dikika baby skull was introduced to the world in *Nature*.
The September 21, 2006, issue of the magazine carried a striking
photograph of the Dikika skull on its cover, along with the title "A
Child of Her Time." Here was the article paleoanthropologists had
been awaiting, and it did not disappoint. My e-mail was flooded with
excited messages from colleagues, and the popular press also
descended on the announcement with an insatiable zeal.

Following the Ministry of Culture and Tourism directives, Zeray
unveiled his discovery at a press conference convened at the
National Museum of Ethiopia. This occasion was timed to coincide
with the *Nature* publication, and everyone who attended was blown
away by the baby announcement. Babies always bring smiles to
people's faces and the Dikika baby was no exception. A widely published photograph of Zeray holding up the baby's skull next to his
shows him smiling from ear to ear, as if announcing, "It's a girl!"

In newspaper and magazine articles the find has been variously
called "Lucy's Baby," "Lucy's Child," "Little Lucy," or "Lucy's Daughter." Zeray felt that the fossils belonged to Ethiopia and that the
Ethiopians should have the right to name their fossils. At the press
conference, someone in the audience called out, "Selam." Apparently, in unison, all assembled agreed to the proposal.

Selam, the Amharic spelling for the Arabic word *salaam*, means
peace. This resonated with Zeray because the word represents hope
for peace in the Dikika region, where armed conflict continues to
be a way of life for the Afar and Issa peoples. Only minutes later
Zeray revealed that it is also the name of his wife. He was now the

father of two babies, a nine-month-old son and a 3.3-million-year-old daughter.

There were many reasons to celebrate Selam. Among them was the fact that her discoverer was Ethiopian. All previous finds had been made either by foreign scholars or by Ethiopians under the direct control of outsiders. But the Dikika baby came to light as the result of an expedition led, for the very first time, by an Ethiopian. This was deeply satisfying to Zeray. He had grown up in Axum, the seat of the great Axumite Empire, and archaeological research there had stimulated his interest in the deep past. There is something fitting about this because the best-known archaeological adventure film is no doubt *Raiders of the Lost Ark,* and locals believe that the Ark is preserved in Axum today, carefully guarded by Coptic Christian priests.

The Dikika specimen is the most complete early hominid baby thus far discovered. The next oldest specimen of comparable preservation is a 50,000-year-old Neandertal infant from Syria. Unlike the *A. afarensis* baby cranium found at the First Family locality at Hadar and pictured in my 1976 *National Geographic* article, the Dikika specimen includes the mandible as well as many of the postcranial bones. The completeness of the Dikika baby is astonishing in that it preserves the skull with teeth and a brain endocast, both collar bones, the backbone from the neck to the lower back, both shoulder blades, many ribs, both tiny kneecaps, fragments of both lower limbs, a partial foot, some hand bones, and even the delicate hyoid bone in the neck that anchors the tongue and the voice box.

Jonathan Wynn explained that the baby must have been buried very quickly before carnivores got to it, because there are no tooth marks on the bones. And the lack of abrasion suggests it might have been buried with the skin still on it. The skin acted to mummify the specimen, preventing the bones from dispersing.

Unlike the First Family we found at Hadar, whose bones had been interred in a river channel, I learned from Jonathan that the Dikika baby was found in a slow-moving delta channel. Maybe the baby merely slipped into the water and drowned, he mused.

What Zeray and his colleagues have revealed about the baby so far is already expanding our knowledge of the anatomy and behavior of *A. afarensis*. The reason I say "so far" is that the mandible has not yet been removed from the cranium, so parts of its anatomy remain concealed. Nevertheless, Zeray is now convinced that the baby belongs to *A. afarensis*. He reached this conclusion after carefully identifying a list of anatomical features in the baby's skull that are seen in the *A. afarensis* collection from Hadar and other sites. But when diagnosing the species identity of the Dikika baby, he also had to demonstrate how the specimen differed from other closely related species such as *A. africanus*. I went through a similar exercise when I evaluated anatomical characteristics in the face of Selam, Dart's Taung baby, and A.L. 333-105 from Hadar. In the end it was clear that with its narrow nasal opening, vertical chin, and projecting snout, among other characteristics, Selam grouped with *A. afarensis*, not *A. africanus*.

We know that adult *A. afarensis* males and females were sexually dimorphic. But how can Zeray be so convinced that the baby is female, when sexual dimorphism does not characterize children? The answer lies in the size of the permanent, or adult, teeth that Zeray knew would be developing deep inside the upper and lower jaws of the child. Thanks to three-dimensional computed tomography (CT) scanning, Zeray was hopeful that he might catch a glimpse of the internal structure of the specimen, including the developing crowns of the adult teeth.

Unfortunately, the right sort of scanning equipment was not available in Addis Ababa, but Zeray obtained permission from the Ministry of Culture and Tourism to export the precious specimen to another country. Working with Fred Spoor from University College London, an expert in the CT scanning of fossils, Zeray scanned the Dikika baby in neighboring Kenya, at the Diagnostic Centre in Nairobi. The CT machine employs the basic principles of X-rays to take micron-thin slices through a specimen—be it a car accident victim in an ER or a fossil skull—that are stored as two-dimensional images in a computer.

In the virtual image Selam's permanent incisors, canines, pre-molars, and first molars were all clearly visible. Plotting the measurements of these teeth against those of *A. afarensis* adults revealed that Selam's values consistently fell in the lower half of the distribution of all adult *A. afarensis* teeth. In other words, the baby's unerupted adult teeth were small, validating the assumption that the Dikika child was female.

To determine how old Selam was when she died, Zeray compared her teeth to the African ape model of dental development, which suggested she was about three. This corresponds to the time when chimpanzees are weaned. I thought back on the Hadar baby skull and the jaw remains of two other babies at the First Family site and their estimated ages at death. They had very similar states of dental eruption—the two milk molars had erupted and the permanent tooth crowns were still developing in the jaws. I conjectured that perhaps age three was when early *A. afarensis* children began to move out on their own, making it an especially dangerous time for them. Or perhaps it was just coincidence that all three infants died around the same age. In any event, we clearly need to try to scan the Hadar infants, to see if we can replicate Zeray's clever determination of gender.

Fred Spoor has been studying the organ of balance—housed in the bony labyrinth of the inner ear—for several years using CT data. He has imaged more than forty-two different species of primates, including hominids such as South African *Australopithecus* and Neandertals. The bony labyrinth contains sense organs used for the perception of movement and spatial orientation. The organ of balance houses three semicircular loops that sense head movement and let your brain know where your head is located, contributing to your balance and preventing blurred vision. When the sensors in the semicircular canals are not working, say, when someone suffers from Ménières disease, the result is vertigo.

What Spoor's studies showed was most intriguing. He found that in mammals that are very agile and highly mobile, the canals are demonstrably larger than in slower, more cautious creatures. For example, Fred pointed out that the bush baby, a little leaping primate, has much larger semicircular canals than does the slower-moving, nocturnal loris. When examining the bony labyrinth in *Australopithecus*, he noted that the proportions of the canals resembled the condition seen in the great apes, not humans. He concluded that locomotion in *Australopithecus* lacked movements such as running and jumping, both features of modern bipedalism.

When Zeray and Fred examined Selam's inner ear region, the three tiny loops in the labyrinth were clear as a bell on both right and left sides. After carefully measuring the arc size of each loop, they compared their observations with the data Fred had previously collected. The conclusion was unambiguous; the inner ear of the Dikika baby was nearly identical to that seen in *A. africanus.* Because this arrangement is similar to that of the living great apes, they concluded that *A. afarensis* lacked "fast and agile bipedal gaits."

One day in the lab, Zeray was cautiously chipping away at the stone matrix that filled in the space between the right and left halves of the mandible, the place we might call the floor of the mouth, where the tongue is situated. I knew why he was being so careful. One of the most delicate bones in the body is housed there—the hyoid. Fossil hyoids are literally unknown except for one specimen found in a 60,000-year-old Neandertal from Israel known as Kebara 2. But given the Dikika baby's exquisite preservation, Zeray thought it just might contain the hyoid, and he wanted to exercise extreme caution. After I left Addis Ababa I learned through Bill Kimbel that Zeray's cautionary efforts had paid off. Astonishingly, the hyoid was preserved and pushed up against the hard palate.

The bone weighs only grams; it serves as an anchor for the tongue and the voice box and, most important, holds the throat open. This tiny, highly mobile bone does not articulate with other bones but rather is supported by muscles that connect with the tongue. Although the hyoid in the Dikika specimen has not yet been freed of matrix, some anatomy can be discerned. The hyoid is U-shaped, with the body positioned forward and two horns trailing backward that form the sides of the U. The baby's hyoid bears a strong resemblance to that of the African apes, especially in its expanded body and enlarged cavity. In African apes the enlarged cavity is an extension of air sacs in the upper chest that store up to 6 liters of additional air. One interpretation of the function of these air sacs is that they provide additional forced air during vocalizations for communication between different animals.

Zeray has suggested that when Selam was crying out for her mother, she would have sounded more like a chimp than a human. In an accompanying commentary on Zeray's *Nature* paper, the anatomist Bernard Wood, of George Washington University, went even further. Noting the reduced canine size in *A. afarensis,* he suggested that "these air sacs might have been a way in which males established a dominance hierarchy, and females judged the quality of a potential mate." This may very well be pushing the behavioral envelope too far, but Zeray's discovery of an *Australopithecus* hyoid is certain to catalyze a whole new area of research. Why do African apes have air sacs and how are they related to behavior? And what exactly can we infer about a 3.3-million-year-old hyoid?

One of the most provocative observations presented in Zeray's *Nature* article pertains to the rate of brain growth. Using the CT images, his team estimated that the baby had a brain size of around 330 cubic centimeters, about the same as a three-year-old chimpanzee. Adult *A. afarensis* skulls range in size from as small as 375 cubic centimeters to 550 cubic centimeters compared to the brains of adult chimpanzees, which range from 282 to 500 cubic centimeters. Zeray compared rates of brain growth in chimps and *A. afarensis* and pointed out a very intriguing implication. At three years old a chimp has attained more than 90 percent of adult brain size, whereas a three-year-old *A. afarensis* has realized only 64 to 88 percent of adult brain size. This implies that the absolute brain growth rate of *A. afarensis* was slower than that of chimps.

If it's true that we see the beginning of the slowing down of brain growth in *A. afarensis*, it is hugely important for understanding the behavior of these ancient hominids, as Owen Lovejoy remarked on the *Scientific American* blog in a post discussing the Dikika baby. He pointed out that "primates, as well as many other social mammals, learn by experiencing a long period of development while protected by their parents and social systems. It is this experience that provides them the necessary knowledge about the world into which they will soon graduate."

It follows that an extended subadult period for learning a complex set of behavioral strategies useful in adulthood demands a better-equipped brain to store and evaluate that information. Lovejoy concluded that even "a small delay in the development of *A. afarensis* would have 'set the stage' for an eventual increase in brain size encouraged by natural selection." What an elegant way to tease information from an ancient fossil and evaluate it within the evolutionary framework of what we know about primates today.

Bones preserved in the Dikika specimen also promise to add significantly to the debate between the "arborealists" and the "terrestrialists." The former group, led by Jack Stern at Stony Brook University, sees *A. afarensis* as having spent lots of time climbing, whereas the terrestrialists' view, championed by Owen Lovejoy, holds that *A. afarensis* was essentially a ground-walking biped. Selam came with two beautifully preserved shoulder blades, or scapulas. Lucy, for comparison, had only a fragment of the right shoulder blade, the concavity called the glenoid fossa, where the round head of the upper arm bone, the humerus, articulates. In arboreal apes the fossa points upward and in humans it faces directly sideways. Assessing the orientation of the fossa in Lucy was problematic because of its fragmentary nature, but the arborealists were convinced it faced upward as in arboreal apes. The terrestrialists, for their part, concluded that as in modern humans, it faced out to the side. The Dikika scapula settled the argument that the glenoid fossa faces cranially, or upward, as it does in apes.

The shape of the scapula is also rather intriguing. The overall shape, that is to say, the height and breadth of the baby's scapula, is more similar to that of humans and gorillas than to the vertically elongated chimpanzee scapula. I think I may know why. The broader shoulder blade of Selam, gorillas, and modern humans may represent the ancestral condition that was retained in gorillas and us, as well as in *A. afarensis*. The elongated chimpanzee shoulder blade resembles that of other primates, like monkeys, that are mostly arboreal. This is an important difference. We know that chimps are much more arboreal than gorillas. And the anatomy of the chimp's scapula is consistent with a more arboreal life, compared to that of the much larger, more terrestrial gorilla. One more aspect of scapula anatomy deserves mention. In gorillas and chimps, the scapular spine that divides the bone into upper and lower parts has a fairly vertical orientation. In the Dikika baby and especially humans, the spine is more horizontal. Muscles that attach above the spine are used for raising the arm in climbing and appear large in tree-climbing apes, very reduced in humans, and intermediate in the Dikika baby. These two features, spine orientation and muscle size, argue for a reduced role for the arms in climbing.

What does all this mean for the *A. afarensis* mode of locomotion? Stern has seized on the baby's curved finger bones and upward-facing glenoid as confirmation of his theory that *A. afarensis* spent a lot of time in the trees. But Lovejoy continues to view these features in the way that we might interpret our appendix: as remnants from our past, evolutionary baggage left over from a time when our ancestors were accomplished arborealists. I tend to agree with Owen on this. Evolution is a slow process and not all anatomical regions of a body evolve at the same pace. Bipedalism appears to have been a very early behavior that distinguished us from the quadrupedal apes, and apelike features of the upper body began to morph only after our ancestors became dedicated bipeds.

PART 2

Lucy's Ancestors

Hadar female *A. afarensis* skull (*left*), 3.1 million years old (*photograph by Bill Kimbel*), and a male *A. afarensis* skull, 3.0 million years old (*photograph by Don Johanson*).

Reconstruction of a Neandertal group in a rock shelter.
(*Illustration by Michael Hagelberg*)

Above left: Reconstruction of a male and female *A. afarensis*.
(*Illustration by Michael Hagelberg,* ASU Research Magazine)

Above right: Bill Kimbel reassembles fragments of the A.L. 444 male
A. afarensis skull in the paleoanthropology laboratory at the National
Museum of Ethiopia. (© *Enrico Ferorelli*)

Don Johanson working on the A.L.
438 ulna at Hadar. (*Courtesy of the
Institute of Human Origins*)

Yoel Rak (*left*), Don Johanson, and
Bill Kimbel discuss new hominid
finds at Camp Hadar in 1993.
(*Photograph by Nanci Kahn/Institute
of Human Origins*)

Camp Hadar and Awash River with Kenia Koma in the background.
(*Photograph by Don Johanson*)

Excavation at the first family locality (A.L. 333). The thin green layer at the top of the ridge in the foreground is the 3.24-million-year-old Triple Tuff.
(*Photograph by Don Johanson*)

Hadar Research Project surveyors on the outcrops look for fossils. (*Photograph by Don Johanson*)

Hadar Afar guards (*from left*) Hamadu Meter, Dato Adan, Edris Ahmed, and Abdu Mohamed. (© *Enrico Ferorelli*)

Hadar Research Project team member Humed Waleno reads from the Koran at Hadar. (*Photograph by Don Johanson*)

Hadar Research Project team, 1994. (*Photograph by Don Johanson*)

Two hand bones, the second and third meta-carpals from A.L. 438, in their anatomical position. (*Photograph by Don Johanson*)

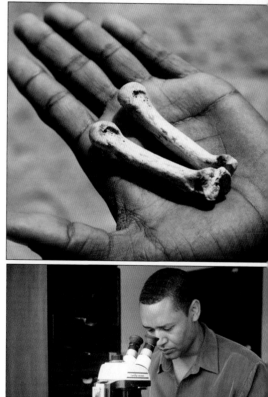

Zeresenay Alemseged removing stone matrix from the 3.3-million-year-old Dikika baby at the National Museum of Ethiopia. (*Photograph by Don Johanson*)

Hadar Research Project surveyors returning to camp at sunset. (*Photograph by Don Johanson*)

Excavation and screening at A.L. 444, the site of discovery of the male *A. afarensis* skull. (*Photograph by Don Johanson*)

Omar Abdulla and Elizabeth Harmon mapping at Hadar. (*Photograph by Don Johanson*)

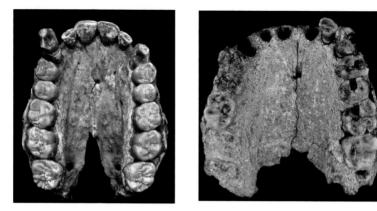

Two upper jaws from Hadar. The palate on the left is long from front to back, narrow from side to side, and the tooth rows are parallel—all typical of *A. afarensis*. The palate on the right is short from front to back, wide from side to side, and the teeth are set in a parabolic arch characteristic of the genus *Homo*. (*Photographs by Don Johanson*)

Sahelanthropus tchadensis, a 7-million-year-old cranium from Chad. (*Courtesy of Michel Brunet*)

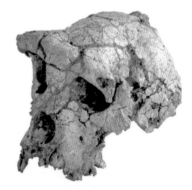

Kenyanthropus platyops, a 3.5-million-year-old cranium from northern Kenya. (*Photograph by Fred Spoor, courtesy of the National Museums of Kenya*)

Right: The 18,000-year-old skull of *Homo floresiensis,* a diminutive hominid from Indonesia also known as "the hobbit." (© *Peter Brown*)

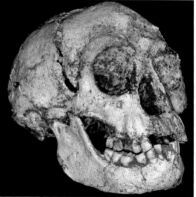

The Dikika baby encased in sandstone prior to cleaning. (*Photograph by Don Johanson*)

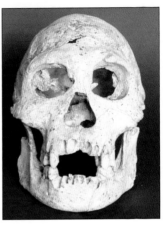

One of the 1.8-million-year-old skulls of *H. ergaster*, D2700, from Dmanisi in the Republic of Georgia. (*Courtesy of David Lordkipanidze*)

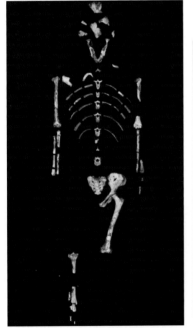

The Lucy skeleton, A.L. 288-1, known by the Afar as *Heelomali*, meaning "she is special." (*Courtesy of the Institute of Human Origins*)

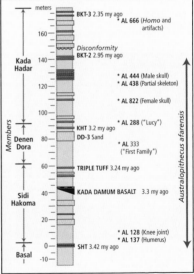

The Hadar Formation.

The Dawn of Humankind

Where, then, must we look for primaeval Man? Was the oldest
Homo sapiens *pliocene or miocene, or yet more ancient? In still*
older strata do the fossilized bones of an Ape more anthropoid, or
a Man more pithecoid, than any yet known await the researches of
some unborn paleontologist?
— THOMAS HENRY HUXLEY, *EVIDENCE AS*
TO MAN'S PLACE IN NATURE, 1863

One of my fondest childhood memories is of reading Thomas
Henry Huxley's remarkable book, *Evidence as to Man's Place in
Nature,* as a teenager. It was then that I became enthralled with
human evolution and the notion that we and apes share a common
ancestor. Huxley's book picked up where Charles Darwin's *On the
Origin of Species* left off. Mild-mannered Darwin, knowing that his revolutionary theory of evolution by natural selection would ruffle the
feathers of Victorian establishment—and especially his deeply reli-
gious wife, Emma—left the subject of human origins all but out of
his book, writing only that "light will be thrown on the origin of man
and his history." That statement alone was enough to ignite a
firestorm of controversy at a time when scientists and nonscientists
alike were governed by the idea that God created all organisms in
their current form, paying special attention to humans. But Huxley,

often described as "Darwin's bulldog," met the opposition head-on. Surveying the fossil remains of primates and humans, he argued for evolutionary continuity between the apes and us. He further concluded that humans were more closely related to the African apes than the Asian apes.

In 1871, perhaps emboldened by Huxley's efforts, Darwin produced his second-most important work, *The Descent of Man.* In it he deemed the chimpanzee and gorilla our closest living relatives based on anatomical similarities between us and our ape kin. (Molecular and genetic findings have since bolstered that view and revealed the chimpanzee, specifically, as our closest living relative.) Darwin then predicted that the earliest ancestors of humans would turn up in Africa, where our simian cousins live today. But the fossilized remains of those first hominids have proved extraordinarily elusive.

When I found Lucy in 1974, she was the one of the oldest known human ancestors. Ancient and primitive-looking though she was, however, we knew there had to be even older, more apelike hominids out there. And somewhere, entombed in the strata of deep time, was the last common ancestor of humans and apes. But try as we might, we simply could not definitively trace humanity's roots back beyond 3.6 million years ago. To probe our past beyond *afarensis* was to fall into a wormhole—and emerge on the other side 9 million years earlier, in the Miocene epoch.

The Miocene, which spans the period between 22 million and 5.5 million years ago, has long fascinated paleoanthropologists, for it was a time when Earth really was the planet of the apes. Today only a handful of apes survive—the orangutan and gibbon in Asia, and the chimpanzee, bonobo, and gorilla in Africa—and most of them are hanging in the balance, thanks to human pressures. During the Miocene, however, as many as one hundred ape species flourished throughout the Old World. One of these gave rise to the human lineage.

In the 1960s it was generally believed that the last common ancestor of hominids and apes lived deep in the Miocene, as long as 20 million years ago. This belief hinged on a fossil that had been discovered three decades before in the Siwalik Hills of northern India. G. Edward Lewis, the Yale University graduate student who found the upper jaw fragment, named it *Ramapithecus punjabicus*, after Rama, a beloved hero of Hindu mythology. Lewis attributed the 7- to 8-million-year-old specimen to a hominid. But it wasn't until the sixties that other researchers embraced that idea, persuaded by similarities between it and later hominids. Among *Ramapithecus*'s hominid-like traits were its thick tooth enamel, small canines, an inferred delay in molar eruption indicating a prolonged period of childhood learning, and the presumed V shape of the tooth rows (in contrast to the U-shaped rows in apes). Though there was precious little material to go on, proponents of *Ramapithecus* as a hominid—notably Elwyn Simons and David Pilbeam, both then at Yale—named more than a dozen traits linking *Ramapithecus* to the australopithecines. They even speculated about the creature's behavior, taking a cue from none other than Darwin himself: Darwin postulated that as the canines shrank, humans began to use tools. Considering the small size of the *Ramapithecus* canines, Simons and Pilbeam reasoned, it probably relied on its hands and tools. Furthermore, if the creature was using its hands to make and wield tools, it was, in all likelihood, walking upright.

Ramapithecus enjoyed the distinction of being the earliest putative hominid until the 1970s, when new specimens from Kenya came to light. These fossils revealed that the Siwalik maxilla had been reconstructed incorrectly. In fact, the tooth row was more U-shaped— that is to say, more apelike. That left thick enamel and a small canine as the primary traits linking *Ramapithecus* to humans. But these traits, too, would betray Simons and Pilbeam. Pilbeam renewed the search for *Ramapithecus* specimens in the Siwalik Hills, this time on the Pakistani side of the border, hoping that additional fossils would elucidate the status of *Ramapithecus*. But it so happened that at every site where *Ramapithecus* turned up, they also found remains of another ape with thick enamel, *Sivapithecus*. Eventually, they were left with the inescapable conclusion that *Ramapithecus* had a small canine not because it was a hominid, but because it was a female *Sivapithecus,* an ape believed to be a relative of the modern orangutan. Siva, the Hindu god of destruction, had put to rest the notion of *Ramapithecus* as a hominid.

The dethroning of *Ramapithecus* rattled paleoanthropologists, for it illustrated in no uncertain terms the perils of classifying fragmentary specimens on the basis of just a few dental traits. But it had the happy effect of somewhat reconciling the fossil record of human evolution with the emerging consensus from molecular biologists. Starting in the 1960s, researchers began turning to molecular evidence for insights into human evolution. Studying blood proteins of living apes and humans, they measured the degrees of genetic difference among humans, chimps, gorillas, and orangutans, confirming the assertions of Huxley and Darwin that we are more closely related to the African apes than orangutans. These "molecular anthropologists" then determined that proteins evolve at a steady rate, and that the degree of genetic difference between any two species therefore corresponds to the amount of time that has passed since they last shared an ancestor. That is, by knowing the difference between proteins of two species and the rate at which those proteins evolve, one can theoretically figure out when the species parted ways. Thus, the molecular clock was born.

Now, molecular clocks aren't exactly Swiss timepieces. They need to be calibrated against a divergence date that has been established in the fossil record using radiometric dating methods. The divergence date most commonly used by molecular anthropologists is the one at which Old World monkeys and apes split from a common ancestor, around 30 million years ago. By taking the genetic distance between Old World monkeys and African apes and dividing that by the amount of time that has passed since they diverged, scientists, including Vincent Sarich at the University of California–Berkeley, came up with a rate of evolution that they could then apply to the question of when humans branched off from the African apes. Their answer: no more than 8 million years ago. Indeed, so confident was Sarich in the molecular clock findings that he declared "one no longer has the option of considering a fossil older than about 8 million years as a hominid no matter what it looks like."

Needless to say, Sarich's cocksure claim irked a number of paleo-anthropologists, most of whom would like to think that they can recognize a hominid when they see one (*Ramapithecus* notwithstanding) and that morphology is more reliable than molecules when it comes to identifying hominids. Nevertheless, it's worth noting that no potential hominid remains older than 8 million years have been discovered to date.

Hominids predating Lucy and her *afarensis* kin have surfaced, however. In 1994 Tim White's group announced its discovery of a hominid that would come to be known as *Ardipithecus ramidus* (*ardi* means "ground" in Afar; *ramid* means "root") at a site in Ethiopia called Aramis, around 140 miles northeast of Addis Ababa. At 4.4 million years old, it was a good 1.2 million years older than Lucy and half a million years older than the oldest representatives of her kind. Forty-three specimens representing several individuals have been recovered, including much of a skeleton. Few details have been released thus far. But we know that the creature was quite apelike in having thin dental enamel, strongly built arm bones, and long, curved fingers. *Ardipithecus*'s hominid status derives from the observation that its canines are smaller and less pointed than ape canines and lack the honing complex that apes have, wherein the upper canine sharpens its back edge against the cheek side of the lower third premolar. And the large hole in the underside of the skull called the foramen magnum, through which the spinal cord passes, is located fairly far forward—a position that is typically associated with uprightness, but not necessarily bipedal walking.

Definitive evidence that *Ardipithecus* was a biped has yet to be revealed, however. In the case of Lucy, a multitude of traits in her pelvis, knee, and ankle identified her as a bona fide biped—one who walked very much as we do when on the ground. The pelvis, lower limb, and foot bones that have been recovered for *Ardipithecus ramidus* will be critical in assessing this hominid's mode of locomotion. So far White has said only that the gait of *Ard. ramidus* was unlike that of any known primate, living or extinct, and that, as he told a reporter at *Science*, it is the Rosetta stone for understanding how humans came to walk on two legs.

With a teaser like that it's no wonder that scholars of human evolution are frustrated that the only published account of *Ard. ramidus* remains the preliminary 1994 *Nature* article. Having to wait some seventeen years after the initial find in 1992 for a detailed analysis is just not acceptable to many researchers. The problem is that unlike Lucy, whose bones were in relatively good shape when I found them, the *ramidus* skeleton is an incredibly fragile specimen encased in rock—a nightmare to clean, piece together, and analyze. The skull, in particular, is so badly crushed it has been described as "roadkill." To assemble the pieces, the team is using a virtual reconstruction technique, in which they obtain high-resolution CT scans of the fossil, feed that data into a computer program, and then reconstruct the skull on-screen. Rapid prototyping technology can be used to "print" a three-dimensional model of their completed reconstruction.

All of this work is incredibly time-consuming. Moreover, Tim is a very exacting scientist who is not about to be pressured into saying more about *ramidus* until he is good and ready. But his unwillingness to share more information about the fossils—not to mention access to the remains themselves—in a more timely way has drawn criticism. (So secretive are he and his team about the fossil that it has been referred to as the Manhattan Project of paleoanthropology.) In fact, spurred in part by Tim's actions, some researchers have even proposed that funding agencies such as the National Science Foundation establish a limitation on how long the discoverer of a fossil has exclusive access to that material before having to share it with other investigators.

To be sure, this is a complex issue. On the flip side, researchers can come under fire for publishing too hastily. When Zeresenay Alemseged, the young Ethiopian scholar who found the "Lucy's baby" fossil, published his initial report—more than five years after extracting the specimen from its sandstone tomb—critics complained that he had rushed into publishing without having exposed enough of the fossil first.

I, for one, eagerly await the release of further details about *ramidus.* It will be especially fascinating to see how its pelvis differs from Lucy's and to ascertain how these differences shed light on the evolution of bipedal walking. A skeleton such as the one from Aramis should offer up a plethora of insights into the anatomy of a hominid species 1.2 million years older than Lucy. We already know the thinly enameled teeth are more apelike than those of *A. afarensis,* and researchers predict the rest of the anatomy will be more primitive as well. But it will be particularly interesting to know what specific anatomical regions in *Ard. ramidus* are significantly more apelike than in *A. afarensis.*

Ardipithecus ramidus pushed human origins back 1.2 million years—and showed us that hominids can look a lot more like apes than even Lucy did. But the molecular studies hinted that even older hominids remained to be discovered. Seven years would pass before fossil hunters found challengers to the *Ard. ramidus* throne, however.

In the spring of 2001, maverick paleontologists Martin Pickford and Brigitte Senut of the National Museum of Natural History in Paris declared that they had recovered some 6-million-year-old remains that they believe are those of a hominid. Having found the fossils in northern Kenya's Tugen Hills the prior year, they christened their find *Orrorin tugenensis*—playing off the local Tugen word *orrorin*, which means "original man" and happens to sound like *aurore,* the French word for "dawn." Its nickname was Millennium Man.

The pair has amassed a number of fossils belonging to *Orrorin,* including fragmentary femurs (thighbones), a partial lower jaw, a finger bone, a humerus (upper arm bone), and several isolated teeth. Like *Ardipithecus, Orrorin* has apelike arms and fingers that would be quite suitable for life in the trees, and canine teeth that are large and pointed relative to our own. But Pickford and Senut note that *Orrorin* has smaller molars and larger postcranial bones than Lucy, and is thus more humanlike than she is in those regards.

Despite the great antiquity of *Orrorin,* Pickford and Senut contend that it is more advanced than Lucy, which is half its age. They suggest that *A. afarensis* is a dead-end branch on the human family tree and take the contrarian view that *Orrorin* gave rise to *Homo.* As for *Ardipithecus,* Pickford and Senut argue that it's a chimpanzee ancestor.

Martin and Brigitte have been longtime acquaintances of mine, and they offered me the opportunity to examine the original *Orrorin* material during a visit to Nairobi. The Tugen Hills fossils, under the care of the Community Museums of Kenya, were retrieved from a safety deposit box and shown to me in a museum office. Relatively few scholars had seen the original remains, and I was thrilled to be among the first outside of the project to be given the privilege of studying them.

I was aware that the case for *Orrorin* being a hominid in the first place rested primarily on the assertion that it exhibits the earliest explicit evidence of upright walking—considered to be the defining feature of humankind. Key to this argument are the femurs. The most complete of these—and hence, the one that has received the greatest scrutiny—is a specimen known as BAR 1002'00, which preserves the head and two-thirds of the shaft of the femur. The bottom end of the bone, which forms part of the knee joint, is missing—the work of a hungry carnivore that gnawed it off some 6 million years

ago. This is a shame, because that part of the femur is highly informative when it comes to figuring out whether a creature walked
upright or not. The *A. afarensis* knee joint I discovered in 1973, the
first hominid fossil to be found at Hadar and indeed the first ever to
be found in the Afar, preserved the bottom end of the femur, with its
telltale bipedal traits. But I knew the upper end of the femur held
clues, too.

Several features of *Orrorin*'s femur align it with that of a hominid.
All proper bipeds have a groove on the back surface of the femoral
neck—the length of bone that connects the head to the shaft—where
a muscle known as the obturator externus presses during upright
walking. Running my thumb on the back of *Orrorin*'s femoral neck, I
could feel a broad obturator externus groove. What also caught my
eye was the long femoral neck, which would have placed the shaft at
an angle relative to the lower leg, thus stabilizing the hip during two-
legged locomotion. For comparison, there is no obturator externus
groove on a chimpanzee's femur. And it has a short femoral neck,
which results in their having to resort to an awkward, bowlegged walk
when upright. Lucy, for her part, is indistinguishable from modern
humans in these characteristics. Unlike a *Homo* femur, an *Orrorin*
femur has a small head relative to the broad shaft and a landmark on
the top surface of the bone known as the intertrochanteric line, which
makes it resemble that of an australopithecine.

After my observations of *Orrorin* there was no doubt in my mind
that it had walked upright. Other scholars were still skeptical. But
there was one feature that could convince them: the internal
anatomy of the femoral neck. Scientists agree that the cross-sectional
anatomy of the neck of the femur is critical for distinguishing between bipeds and quadrupeds. During upright locomotion, the
femur head serves to transfer the weight of the upper body into the
lower limb. In response to the stress this loading imparts, the cortical
bone in the lower portion of the neck—the part most vulnerable to
bending forces, and thus breakage—thickens. (The cortical bone

that makes up the upper portion of the neck, which endures relatively little compression, remains thin.) Animals that travel on all fours, in contrast, have an even distribution of cortical bone across the femoral neck.

Although Pickford and Senut see a humanlike pattern in the cross section of the femoral neck, others do not share this view. Owen Lovejoy, a leading authority on bipedalism, has argued that their published CT scans reveal a chimplike pattern of cortical bone distribution, not a human one. The problem is that the CT scans are not especially good and that the upper portion of the femoral neck is eroded or chewed, thereby obscuring the thickness of the cortical bone in that area. Given the differences of opinion over what the CT scans show, Lovejoy and others—notably Tim White, with whom Lovejoy had recently conducted a definitive analysis of a 3.4-million-year-old australopithecine femur from Maka, Ethiopia—called for Pickford and Senut to obtain higher-quality CT scans of the fossil. White also suggested taking X-rays or photos of the inside of the femoral neck. The latter was feasible, because it just so happened that one of the spots where the bone had broken was at the neck. Pickford and Senut would merely have to unglue the fossil to photograph the neck in cross section. They have steadfastly refused to do that, however, claiming that surface details, such as the obturator externus groove, already establish *Orrorin* as a biped, so ungluing the fossil is not justified given how fragile it is.

In March 2008 Brian Richmond of George Washington University and William Jungers of Stony Brook University published the results of their own study of *Orrorin*'s thighbones. Writing in the journal *Science*, they concluded that the pattern of anatomy seen in *Orrorin*'s femur is not only consistent with bipedalism but essentially identical to the pattern evident in Lucy's femur and that of all other australopithecine species. But they refuted Pickford and Senut's argument that *Orrorin* was more humanlike than Lucy.

* * *

Just months after the *Orrorin* bombshell dropped in 2001, another candidate for the title of "oldest hominid" emerged. Yohannes Haile-Selassie, then a Berkeley graduate student under Tim White, reported in the journal *Nature* that new finds in Ethiopia's Middle Awash extended the bloodlines of *Ardipithecus* back as far as 5.8 million years ago. At several sites in this arid, rocky region of the Afar rift, Haile-Selassie and his coworkers had discovered eleven specimens representing at least five individuals that they thought belonged to another, older subspecies of *Ardipithecus ramidus* that they named *Ard. ramidus kadabba*. Since then, they have unearthed half a dozen more teeth and redesignated the hominid a distinct species, *Ard. kadabba* (the species name means "basal family ancestor" in the Afar language). Haile-Selassie, now the curator and head of physical anthropology at the Cleveland Museum of Natural History (my old job), has suggested that this chimp-size creature represents "the first species on the human branch of the family tree just after the evolutionary split between lines leading to modern chimpanzees and humans."

Like *Orrorin*, *Ard. kadabba* has been held up as an early biped. However, the hypothesis that *Ard. kadabba* traveled on two legs hinges on a single left foot phalanx, or toe bone, that has been consigned to this species. The bone's joint tilts upward like a human's rather than downward like a chimp's—a configuration that enables humans to "toe off" when walking. This, in Haile-Selassie's view, surely means that *Ard. kadabba* walked upright. His assessment has won the blessing of Lovejoy, who has studied the fossil. But other researchers question just how humanlike it is. David Begun of the University of Toronto, whose research focuses on Miocene apes, pointed out in an editorial in the March 5, 2004, issue of *Science* that the same joint configuration is evident in the toe bones of *Sivapithecus*, the Miocene orangutan relative, which was certainly not bipedal.

Another serious problem with *Ard. kadabba* is that it comprises fossils that span a 100-square-kilometer area and 600,000 years of time. That toe bone turned up 20 kilometers east of many of the dental specimens used to diagnose *Ard. kadabba* and is some 400,000 to 600,000 years younger than all the other *Ard. kadabba* fossils. There is absolutely no way one can confidently assert that these specimens come from the same species.

The question of whether *Ard. kadabba* was a biped can only be answered through the discovery of more fossils of this enigmatic creature. But the teeth do hint that it might be a hominid—regardless of whether it walked upright. All apes have large, dagger-like canine teeth that are quite fearsome to behold. Because the upper canine hones against the lower third premolar, it maintains a razor-sharp back edge. This "honing canine-premolar complex," as it's termed, is especially pronounced in males, who use their canines to compete with one another for access to females. *Australopithecus* and *Homo,* on the other hand, have small canines that occlude tip to tip and are utterly wimpy by comparison. Scientists believe that hominids lost those fighting teeth as males began to clash less and cooperate more. *Ard. kadabba* lies between the ape and human conditions. An apelike honing process is evident, but the overall size of the canine is somewhat reduced. Yohannes Haile-Selassie contends that this quasi-hominid condition is sufficient to place *Ard. kadabba* firmly on the hominid line, and maybe it is. But on the other hand, it might be prudent to await more definitive evidence.

They say good things come in threes. Just when it seemed likely that paleoanthropologists were going to have to wait another decade for more early hominid discoveries, a third fossil rose to candidacy for the title of first human. Merely a year after Haile-Selassie's initial report on *Ard. kadabba* appeared, the French fossil hunter Michel Brunet of the University of Poitiers unveiled an astonishingly complete skull from northern Chad's Djurab Desert that is around 7 million years old. He and his team named it *Sahelanthropus tchadensis,* which translates approximately to "Sahel hominid from Chad."

Nicknamed Toumaï, or "hope of life" in the local Goran language, the find—trumpeted on the cover of the July 11, 2002, issue of *Nature*—electrified the paleoanthropology community.

Where the skull was found is nearly as important as the fossil itself, because it suggests that humanity may have originated in a different locale than has been postulated. In the 1980s, the paleontologist Yves Coppens, formerly of the Collège de France, proposed that East Africa was the birthplace of humankind. Observing that the oldest hominid fossils then known came from East Africa, and that no chimp or gorilla fossils had been found there, Coppens hypothesized that the Great Rift Valley—a 6,000-kilometer scar stretching from Lebanon in the north to Mozambique in the south—divided a single ancestral African ape species into two populations. The group in the east, he argued, begat humans; the one in the west gave rise to today's apes. It was a tidy story, but experts had recognized for some time the possibility that what appeared to be a geographic separation could instead be an artifact of the incomplete fossil record.

Fieldwork conditions are never easy in Africa. There are always diseases, dangerous animals, and political unrest to contend with. But East Africa, with its natural rock exposures and relatively tolerable climate, has long been the preferred hunting ground of paleoanthropologists. Few have dared venture beyond that comfort zone. Coppens was one of the first to comb the sands of the Djurab for ancient bones, making his inaugural expedition there in 1960. Ironically, it was he who located the fossil beds in which Brunet would eventually discover *Sahelanthropus*. Coppens never found any hominids there himself, save for a partial skull he named *Tchadanthropus uxoris* ("Chad man found by my wife") that was later shown to be a heavily eroded modern human. By 1967 a brutal civil war had erupted, and he could no longer work there safely. He turned his attention to Ethiopia, coleading expeditions to the Omo valley.

It would take Brunet, with his peerless perseverance, to pick up where Coppens had left off in the Djurab. There daytime temperatures routinely soar upward of 120 degrees Fahrenheit, windstorms relegate visitors to their tents for days at a time, and snakes, scorpions, and land mines threaten at every turn. Brunet was no stranger to perilous field-work. He had worked in Afghanistan in the 1970s and narrowly escaped fire from Russian fighter jets, and he had lost a dear friend to malaria during a 1989 expedition to Cameroon. But he had never found a hominid. Brunet's luck changed in 1995 during his second field season in Chad, when his team discovered the partial jaw of an australopithecine that had lived between 3 million and 3.5 million years ago. He assigned it to a new species, *Australopithecus bahrelghazali,* for the Bahr el Ghazal ("river of the gazelles") region in which it was found, though most experts think it belongs to *A. afarensis.* It was an exciting find because it showed for the first time that early hominids lived west of the Rift Valley.

Six years would pass before the discovery of *Sahelanthropus* raised the possibility that the human lineage actually originated out-side of East Africa. The find came on July 19, 2001, when Ahounta Djimdoumalbaye, a Chadian undergraduate on Brunet's expedi-tion, spotted the fossil—a coconut-size skull, its left side severely crushed and its lower jaw missing, but otherwise in remarkably good condition—at a site called Toros-Menalla that Coppens had discov-ered back in the 1960s. A partial lower jaw and a few isolated teeth belonging to the same species would follow. Based on the animal fos-sils in the vicinity of the skull—especially those of elephants, pigs, horses, and antelopes—the team estimated that the creature lived more than 6 million years ago, on the shores of the then-enormous Lake Chad.

The skull is so primitive that Brunet has remarked that *Sahelan-thropus* could virtually touch the point at which our lineage and the one leading to chimps diverged. To be sure, its small cranial capacity (an estimated 360 to 370 cubic centimeters), massive brow, widely spaced eye sockets, nearly absent forehead, and features of the base and rear of the skull are reminiscent of apes. But in other regards,

Sahelanthropus resembles a hominid. Its face lacks the extreme snou-
tiness associated with apes and its teeth have relatively thick enamel.
The canine tooth is especially important in the diagnosis of this individual. It is relatively small (though within the size range of variation
seen in modern African apes), it does not project like ape canines
do, and it is worn at the tip—unlike ape canines, which remain
pointed for life.

No postcranial bones turned up with Toumaï, but hints to its
preferred mode of locomotion may reside in the skull. In apes, the
hole in the base of the skull through which the spinal cord passes—
a feature known as the foramen magnum—is located posteriorly,
where it faces backward. This allows the animal to position its head
to look forward during quadrupedal locomotion. In humans, by
contrast, the foramen magnum sits much farther forward on the
skull base, such that the skull balances over a vertical spine during
upright walking. Toumaï's foramen magnum appears to reside fairly
far forward. This is not ironclad evidence of bipedalism, because
there is some overlap between chimps and hominids in this regard.
But it is suggestive.

The claim that Toumaï is a hominid has not gone unchallenged, however. In a letter to the editor in response to the 2002
Nature paper announcing the find, Milford Wolpoff of the University
of Michigan and his colleagues—including *Orrorin* discoverers Pickford and Senut—charged that it was an ape. The thick brow and features of the base and rear of the skull are reminiscent of an ape that
walks on all fours and dines on tough foods, they argued. As for the
small canine, they attributed that to Toumaï being a female ape
rather than a male human ancestor. Lacking unequivocal proof that
the animal was bipedal, they concluded, the case for Toumaï being a
hominid doesn't hold water. (Pickford and Senut further postulated
that Brunet's find is specifically a gorilla ancestor.)

A riled Brunet responded by comparing his critics to the ones
the great Raymond Dart faced in 1925, when he unveiled a skull of

Australopithecus africanus—the first fossil evidence of an early human precursor from Africa—only to be told (incorrectly) by detractors that the skull was that of a young gorilla. Toumaï's apelike traits were merely evolutionary baggage handed down from its own ape forebear, and therefore had no bearing on the question of the creature's relationship to humans, Brunet countered.

Wolpoff and his colleagues were not to be deterred, however. In 2006 they published a paper in the electronic journal *Paleoanthropology* in which they expanded their argument for *Sahelanthropus* being an ape, not a hominid. The paper warned of the misguided hominid claims that have been made for a number of extinct apes over the years, *Ramapithecus* being the most infamous example. Surveying a bunch of living and extinct apes, the team concluded that characteristics used to establish *Sahelanthropus* as a hominid—notably, that this was a male with small canines that probably made a habit of walking upright—did not stand up to scrutiny. The fossil's sex is indeterminate, the canines are neither large enough to be from a male nor small enough to be from a female, and characteristics of the base and rear of the skull dispute the idea that the animal habitually held its head in an upright position over the spine—a prerequisite for bipedal walking. "I don't think there's a chance *Sahelanthropus* is a hominid," Wolpoff declares. Based on the available evidence, he thinks there's a very real possibility that the direct ancestor of Lucy's species—the 4-million-year-old *Australopithecus anamensis*—is still the earliest definitive hominid on record.

For his part, Brunet has stood by his interpretation. Better evidence of bipedalism would verify his claim. The veteran fossil hunter has promised that more bones will come, though none have been announced since his initial publication nearly seven years ago. In the meantime, clues could arise from analyses of the skull's inner ear. The semicircular canals, which are important for maintaining balance, tend to be smaller in apes and other quadrupeds, whereas they are significantly larger in humans. If researchers can glimpse

Toumaï's semicircular canals using CT imaging, they just might get a better handle on how it got around. I suspect that Toumaï's inner ear labyrinth will very closely resemble that of the African apes.

Of course, there can be only one first hominid that is the ancestor to later hominids. Maybe bipedalism arose more than once, but only one species left descendants. The debate over *Orrorin, Ard. kadabba,* or *Sahelanthropus* stems from the fact that researchers disagree over what makes humans unique. There was a time when bipedalism was the defining characteristic of our lineage. But it may be that a subtler change—the transformation of the canine, for example—preceded that shift and is therefore the trait that first set us apart from our ape brethren. What makes it so hard to assess which, if any, of these contenders is the ur-hominid is the fact that they are represented by mostly different parts, making comparisons among them difficult at best. For the moment it appears that the earliest definitive biped, *Orrorin,* may have also possessed a modified canine. If it did, we still don't know what came first: the transformation of the canine or the appearance of bipedalism. For my money, with the apelike honing canine-premolar complex still present in *Ard. kadabba,* I am betting that bipedalism was the initial change that launched the human career.

The First Australopithecines

Meave Leakey laid the gray fossil casts she had brought with her from Nairobi one by one in a wooden tray lined with bubble wrap. Across the table from her, Bill Kimbel arranged casts of Lucy's bones and those of other *Australopithecus afarensis* individuals for comparison. It was an early Saturday morning in the spring of 2001 and we had gathered in the laboratory of the Institute of Human Origins to try to answer a key question: Could Meave's Kenyan fossils represent the long-lost direct ancestor of *A. afarensis*?

I have known Meave since the early 1970s when she married Richard Leakey. Soft-spoken and a bit reticent, Meave is the polar opposite of Richard, and I enjoyed spending time with her in Nairobi. But since the 1980s, when Richard and I had our falling-out over details of the human family tree, Meave and I had not seen much of each other except at professional meetings. So I was delighted when she accepted an invitation to lecture at Arizona State University, where I had moved the institute in 1997. It would be the perfect opportunity to catch up. Bill and I had seen the fossils that she and her longtime collaborator Alan Walker of Pennsylvania State University had collected from northwestern Kenya when we visited Nairobi a few years earlier, and we were eager to see how they compared with our material from Hadar.

In the early 1980s, expeditions run by the National Museums of Kenya to the Turkana basin started recovering hominid remains almost 4 million years old. The fossils were important because they filled a gap in the fossil record between *A. afarensis* and the last common ancestor of chimpanzees and humans. But frustratingly, the scrappy nature of the remains—most were isolated teeth—made it impossible to attribute them to a particular species. As Meave and Alan recollected in an article for *Scientific American,* "very little could be said about them except that they resembled remains of *afarensis* from Laetoli," the site of the famed hominid footprint trail.

Finding more diagnostic fossils of this 4-million-year-old hominid was critical. Paleoanthropologists already knew a considerable amount about *A. afarensis* and its descendants thanks to a string of discoveries going as far back as 1925, when the anatomist Raymond Dart of the University of the Witwatersrand in Johannesburg placed the fossilized child's skull he had found at the South African site of Taung in a new taxon that he called *Australopithecus africanus.* In the years that followed, Dart and others turned up more examples of *A. africanus,* as well as another type, *A. robustus,* in South Africa. The dating of the South African cave sites that yielded these australopithecines is a little challenging because there are no volcanic ashes or other deposits that can be dated using radiometric techniques. Researchers have thus had to rely on the presence or absence of mammalian species that are known to have lived at a certain time to create a chronological sequence for the sites. This approach reveals that *A. africanus* lived from about 2.8 to 2.5 million years ago; *A. robustus* lived between roughly 2 and 1.5 million years ago.

Both of these species were upright-walking creatures with relatively small brains and canine teeth that were greatly reduced compared to those of their ape kin. But whereas *A. africanus,* like *A. afarensis,* probably had a broad diet, *A. robustus* exhibited a number of adaptations for processing tough-to-eat foods. The exact nature of the diet has never been determined; it evidently required lots of crushing and grinding. Scholars have suggested such menu items as bulbs, roots, nuts, tubers, and even large quantities of small food items like seeds. We know the emphasis was on crushing and grinding because the cheek teeth (molars and premolars) have very large surface areas and hyperthick enamel to extend the life of the tooth. Yet the incisors and canines, which are used for piercing and slicing, are very small considering the large size of the jaws. Subsequent searching in the 1950s and '60s turned up more species with these chewing adaptations and together they became known as the robust australopithecines. They were about the same size as other australopithecines, but their teeth, jaws, and chewing muscles were absolutely massive. This was an adaptation for processing lots of low-quality food, the Cuisinart of human evolution.

Louis and Mary Leakey's discovery of a cranium of one of these robusts put Tanzania's Olduvai Gorge on the map in 1959. The 1.8-million-year-old skull—which would come to be called *A. boisei*—revealed a chewing machine like no other. It had a massive face, gigantic molars, and muscle scars that could only have been created by enormous chewing muscles capable of creating very high bite forces. The muscles anchored to the cheekbones, the masseters, were so large they actually caused the cheekbones to project forward, leaving the center of the face dish-shaped. Meanwhile, a bony ridge along the top of the skull, known as a sagittal crest, would have anchored the other major chewing muscles, the temporalis muscles—the ones you feel in the temporal region on the sides of your skull when you clench your teeth—indicating that they, too, were huge. And its teeth show profound wear and tear: Chips of enamel are missing and the molar surfaces are deeply pitted and scratched from intensive mastication. Exactly what *A. boisei* and the other robusts were eating is uncertain, but various kinds of nuts, seeds, roots, and tubers may have been menu items.

Studies of stable carbon isotopes extracted from hominid tooth enamel are throwing light on this question. Paleoanthropologists expected that such evidence would show that the robusts favored foods from trees and shrubs—so-called C3 plants, which prevail in forested environments. But in 2007 Matt Sponheimer of the University of Colorado at Boulder and Julia Lee-Thorpe of the University of Bradford reported in the *Handbook of Paleoanthropology* that their analysis of thirty-seven samples from South African robusts revealed that 35 to 40 percent of their diet was composed of C4 foods such as grasses and sedges (or animals that ate C4 plants). This apparent increase in dietary breadth may have had important consequences for foraging and social behaviors. For example, exploiting the subterranean portions of grasses and sedges demands use of rudimentary digging implements. Thus far, no *afarensis* or *anamensis* teeth have undergone stable isotope analysis. But hopefully this will happen in the near future. If so, we may find out that *anamensis* was the first hominid to venture out of the forests for food.

Even older examples of robusts have turned up since the discovery of *A. boisei*. In 1985 Alan Walker found a startling fossil west of Lake Turkana known as the Black Skull. So-named for the manganese-rich minerals that stained it bluish black, the 2.5-million-year-old specimen was considered by Walker and his colleagues to be an early representative of *A. boisei*. But when Bill Kimbel, Tim White, and I compared the Black Skull to other species of *Australopithecus*, we concluded that it could not be an *A. boisei*, since it shared only a few distinctive anatomical traits with this species, and in fact shared more similarities with *afarensis*. We suggested placing it into an already named species, *A. aethiopicus*, proposed in 1968 by Yves Coppens and another French scholar Camille Arambourg, for a 2.5-million-year-old robust mandible found in Omo, southern Ethiopia. The evolutionary ramifications of the specimen are fascinating: It now appears that *A. africanus* begat *A. robustus* in southern Africa, and that *A. aethiopicus* gave rise to *A. boisei* in East Africa. Should further discoveries bear this out, the robust australopithecines will stand as a rare example of

parallel evolution, in which members of two descendant lineages of
A. afarensis independently evolved similar adaptations, presumably in
response to environmental change. Their shared strategy served them
well, enabling them to occupy a sizable geographic area and persist
for at least 1.5 million years—an unusually long run. But eventually
the robusts lost out, disappearing around 1.4 million years ago. In so
doing, they ceded the planet to the third of *A. afarensis*'s descendant
lineages, the one that led to *Homo*.

From the rich record of *Australopithecus* available to us by the
1990s, we had a pretty good handle on the biology and environmental niches of its various species. But we had no idea where *A. afarensis*
came from. There simply were not any diagnostic remains older
than 3.6 million years. So when Meave and Alan began finding
hominid jaws and other informative parts that were older than that,
the import of the finds could not be overstated.

The fossils came from two sites: Allia Bay, located on the eastern
shore of Lake Turkana, and Kanapoi, which lies across the lake and
about 145 kilometers south of Allia Bay. At Allia Bay, the team recovered thirty-one specimens—jaws and teeth—from a now-dry channel of an ancient river system that once flowed across the Turkana
area. The site is a bone bed comprising millions of fragments of
tooth and bone from an array of creatures, many of them aquatic,
deposited there by the primeval river. Most of the Allia Bay fossils are
quite weathered, having been transported downstream. But there
are also some well-preserved bones from animals that lived in the
forest that lined the river, hominids among them. Conveniently, a
layer of volcanic ash known as the Moiti Tuff overlies the bone bed.
Radiometric dating places the tuff at just over 3.9 million years old,
and because the fossils lie several meters below the tuff, it is certain
that they predate it.

The work at Kanapoi has yielded even more hominid specimens—forty-seven in all. These are not the first hominid fossils to
surface there, however. In 1965 Bryan Patterson, a Harvard paleon-

tologist, recovered the elbow end of an upper arm bone, or humerus, that seemed to be older than any *Australopithecus* fossils then known. But the fragmentary nature of the find obscured its true significance. There just wasn't enough of it to be able to assign it to a particular species. Three decades would pass before scientists, this time led by Meave, mounted another expedition to remote Kanapoi. Little did they know that it would turn out to be one of Turkana's crown jewels.

The fossils that Meave and her team wrested from Kanapoi's pebble-covered badlands come from sediments deposited by a precursor of the modern-day Kerio River that emptied into an ancient lake known as Lonyumun. Most of the specimens are quite fragmentary, the work of hungry carnivores. But the fossil hunters did find several largely complete jaws and a nearly complete shinbone, or tibia. And here, as at Allia Bay, the fossils lie under a layer of volcanic ash (the Kanapoi Tuff) that provides a minimum age for the remains—in this case, a little more than 4 million years. Using argon-argon dating, the team determined that the oldest hominid specimens at Kanapoi are about 4.2 million years old.

Careful analysis of these bones, along with Patterson's humerus and the material from Allia Bay, persuaded Meave and Alan and their colleagues that they had found a new species of australopithecine. They named it *A. anamensis,* using the Turkana word *anam,* which means "lake," to refer to both Lake Turkana and the primeval Lake Lonyumun. Together, the hard-won *A. anamensis* fossils reveal an ancestor with a mix of primitive and advanced characteristics. The mandible, for example, is apelike in being small and narrow, with perfectly parallel tooth rows that form a U shape instead of widening at the back of the mouth as a human's do. Yet the teeth lack the thin enamel that the African great apes and *Ardipithecus ramidus* have and exhibit the same relatively thick enamel that all australopithecines possess. This feature hints that *A. anamensis* may have already begun adapting to a regimen of much tougher foods.

Meanwhile, perhaps most telling of all, the tibia exhibits a human-like thickening of bone both at the knee and ankle joints that served as a shock absorber during upright walking. What specifically fascinated Bill and me about the bones from Kanapoi and Allia Bay was the possibility of an evolutionary relationship between *A. anamensis* and *A. afarensis*. It was plausible that these geologically older specimens, ranging in age from 4.2 million to 3.9 million years of age, were directly ancestral to Lucy and the gang. Of course, just because one species predates another does not necessarily imply an evolutionary tie between them. We needed to conduct an exhaustive comparative study of the two species to see if *A. anamensis* possessed features more primitive than those of *A. afarensis*.

This is easier said than done, because although we can be certain that species had ancestors, we cannot know whether a particular fossil species left any descendants. But there is some safety in numbers. With a reasonable selection of specimens from Lake Turkana and the burgeoning collection of *A. afarensis,* we could test the idea that there was an ancestor-descendant relationship between the two species. If this null hypothesis held up and was not rejected by our analysis we might see confirmation of Darwin's idea of anagenesis—the emergence of a species along an ancestor-descendant lineage through time without branching—which would be a first in early hominid evolution.

And so it was that Meave, Bill, and I, along with Charlie Lockwood, Yoel Rak, Zeray Alemseged, and Carol Ward, assembled in the lab that Saturday morning to begin the arduous process of precisely comparing the anatomy of *A. anamensis* and *A. afarensis*. After much conversation we agreed on a list of twenty anatomical traits that could be assessed in the samples. Charlie Lockwood walked up to the white board and drew four columns, one each for Kanapoi, Allia Bay, Laetoli, and Hadar. On the left side of the board he listed the anatomical features we would consider, such as the shape of the upper canine and the size of the second incisor. Each of us then

examined that trait in the fossil samples from the four sites. Once
we reached a consensus, we scrawled a brief description in the
appropriate column. At the end of a very long but rewarding day we
all walked over to a restaurant in Tempe called the House of Tricks
and reviewed our findings over dinner and some wonderful caber-
net sauvignon. By the time the meal drew to a close, there was little
doubt in our minds that *A. anamensis* had spawned *A. afarensis,*
and we eventually published our findings in the *Journal of Human
Evolution.*

Additional *A. anamensis* fossils recovered since then have reaf-
firmed this view and fleshed out our understanding of this all-
important species. In 2006 Tim White and his colleagues announced
that they had unearthed thirty-one specimens of *A. anamensis* from
two sites in their Middle Awash study area, Aramis and Asa Issie,
extending the known range of this human ancestor by nearly 1,000
kilometers to the northeast. Most of these 4.1- to 4.2-million-year-old
fossils are teeth, but the team also found hand and foot bones, and
most of a thighbone, all of which are reminiscent of Lucy's. Importantly, both *A. anamensis* and *A. afarensis* have been found in the
Middle Awash, and it is clear that the former precedes the latter,
upholding the proposed ancestor-descendant relationship.

The Middle Awash finds also help to clarify the preferred habitat
of *A. anamensis.* Whereas the *A. anamensis* fossils from Allia Bay and
Kanapoi turned up amid remains of animals from a variety of ecological settings, the ones from the Middle Awash co-occurred exclusively
with monkeys, kudus, and other creatures that live in closed, wooded
environments. This paleoenvironmental evidence, along with comparable evidence from *A. afarensis* sites, has sounded the death knell
for the so-called savanna hypothesis that reigned supreme when I was
a student. According to this model, climate change beginning some
10 million years ago resulted in the replacement of forests with grasslands, where our ancestors had to stand up to see over the tall grass.
I've never found this argument very persuasive. Early hominids were

small and bipedalism is a clumsy, very slow mode of locomotion. It seems to me that standing up and announcing that you are on the menu for some marauding carnivore would have quickly led to our extinction. These latest findings indicate that our primeval predecessors must have become bipedal in the forests before moving out onto the grasslands.

Ironically, the main reason for Meave's visit to ASU was to give a talk about another hominid her team had found recently, one that she hinted could have given rise to *Homo.* The cover of the March 22, 2001, issue of *Nature* displayed the find, a 3.5-million-year-old cranium from a site called Lomekwi in West Turkana. Meave and her colleagues noted that unlike *A. afarensis,* which was the only hominid previously known from that time, their specimen has a flat face. And if the single molar it preserves is any indication, its molars were small compared to the large chewing teeth of *A. afarensis.* At the same time, however, it lacks derived *Homo* traits, such as a large brain. Based on these and other features, the investigators deemed it appropriate to coin a new hominid genus and species name for the skull from Lomekwi, *Kenyanthropus platyops,* the "flat-faced man from Kenya."

As a contemporary of *A. afarensis, Kenyanthropus* was presumably another offshoot of *A. anamensis.* This in and of itself was a revelation. Paleoanthropologists have long known that multiple hominid kinds shared the landscape in later stages of human evolution. Most of us thought that prior to 3 million years ago, however, the various species of *Australopithecus* made up a single evolving lineage. If Lucy's species shared eastern Africa with another type of human, then hominid lineages must have started diversifying earlier than previously thought.

But Meave and her coauthors took their argument one step further, observing that *Kenyanthropus* bears a resemblance to a 1.8-million-year-old early *Homo* cranium known as KNM-ER (Kenya National Museums, East Rudolf) 1470 that Richard Leakey's team had found in 1972 on the eastern shores of Lake Turkana. *Kenyan-thropus* could be ancestral to that hominid, they proposed. The suggestion that *Kenyanthropus* might belong to a lineage that led to *Homo* had monumental implications: If *Kenyanthropus* gave rise to *Homo,* then that would relegate Lucy's species to a dead-end branch on the family tree, effectively ousting *A. afarensis* from our ancestry.

With only a single highly fragmented and distorted cranium of *Kenyanthropus* to go on, I felt it was premature to assert such a revolutionary concept. I think that to draw an evolutionary connection between specimens separated by 1.7 million years is, to say the least, a real stretch. Moreover, I wondered whether the skull actually represented a new taxon. By Meave's own admission, the specimen was a mess when she first laid eyes on it. Grass and tree roots had pushed their way into the skull, which itself had become largely encased in a rocky matrix. Those bits of bone that were visible were riddled with tiny cracks. Meave and her collaborators spent more than a year cleaning, reconstructing, and analyzing the cranium, as well as a partial upper jaw found at the same site. And it undoubtedly looks far better now than it did when it was discovered. But I couldn't help questioning whether we can ever know the true shape of a fossil that suffered so much postmortem damage.

I wasn't the only one with doubts. In an editorial published in *Science* in 2003, Tim White voiced serious concerns about the validity of using such a highly damaged specimen to establish a new taxon, drawing attention to its "expanding matrix distortion," or EMD, in which small pieces of bone are separated by matrix-filled cracks of varying sizes. He noted that EMD has influenced almost all of the traits said to distinguish *Kenyanthropus* from other hominids. It may well be that if the more than 1,100 pieces making up the face alone—all separated from one another by rocky matrix—were meticulously repositioned in their proper alignment, the specimen would be seen to be nothing more than a Kenyan variant of *A. afarensis.*

For her part, Meave stands by her assessment. If she is correct, then paleoanthropologists will have new questions to grapple with. For instance, although *Kenyanthropus* has small cheek teeth, the forward position of the cheekbones that gives the skull its flat face suggests that it had powerful chewing muscles. What accounts for the mismatched traits? Only time and more fossils less distorted than the first will resolve the mystery. In the meantime, most experts are betting that the *afarensis* lineage is the one that led to *Homo*.

There are still gaps to fill if we are to prove the connection between *A. afarensis* and *Homo*, however. Immediately after 3 million years ago, the most recent occurrence of *A. afarensis*, the hominid trail goes cold until around 2.5 million years ago. It was sometime during this 500,000-year interval of time that the *Homo* lineage appeared in East Africa. Was *A. afarensis* itself the direct ancestor of *Homo*, or was there another species of *Australopithecus* intermediate between them? Traditionally, scholars considered *A. africanus* to be just such an intermediate. But in 1979 Tim White and I published a paper in the journal *Science* that concluded otherwise. Our in-depth analysis of the anatomy of the various species of *Australopithecus* showed that *A. africanus* could not be an ancestor to *Homo* simply because the species had traits that were already evolving in the direction of the robusts, such as heavily buttressed jaws and enlarged cheek teeth. We moved *A. africanus* to a side branch that gave rise to *A. robustus*. Raymond Dart was still alive, and even though he was uncomfortable with that notion (after all, he had found and named *A. africanus*) he was gracious and wrote a short foreword to a much more extensive paper on *Australopithecus* that we published in the *South African Journal of Science* in 1981.

A. *africanus*'s own ancestry in South Africa is still pretty murky, though. Present evidence indicates that Lucy's species gave rise to *A. africanus*, but there's a considerable gap in morphology between the two. Fresh insights may come from a remarkable fossil currently being excavated by Ron Clarke of the University of the Witwater-srand in Johannesburg. The specimen is located in the Silberberg Grotto, a cavern located deep within the famous site of Sterkfontein that has yielded hundreds of *A. africanus* fossils. Clarke and his team have been slowly chipping away at the cave deposits since 1997 in an effort to expose what appears to be a nearly complete skeleton of *Australopithecus* that might represent a species ancestral to *A. africanus*. Geological dating of the specimen has been very difficult, but based on paleomagnetic results the team has assigned a very pre-liminary date of 3 million to 3.5 million years to the skeleton. If correct, this would put it in the *A. afarensis* range of time.

A new wrinkle in the story emerged in 1999, when Berhane Asfaw announced that he and his colleagues had found a 2.5-million-year-old cranium at a site called Bouri in the Middle Awash. The specimen contains such a startling mix of anatomical features that they named it *Australopithecus garhi* (*garhi* means "surprise" in the Afar language). Like *A. afarensis, A. garhi* has a small cranial capacity of 450 cubic centimeters, a U-shaped palate and a projecting face, among other shared characteristics. Yet it possesses a canine tooth larger than that of any *Australopithecus* or *Homo* species, as well as giant cheek teeth, like those seen in the later robust australo-pithecines. This bizarre mosaic of traits distinguishes *A. garhi* from all other early hominids.

But the cranium wasn't all the fossil hunters found at Bouri. They also recovered a partial skeleton of comparable antiquity from a spot nearby. Because the skeleton does not include any cranial or dental parts, it cannot be attributed with certainty to *A. garhi*, but if it is a member of that species, it tells us something quite interesting about the evolution of limb proportions. The femur is long compared to the upper arm bone, which is a modern trait associated with efficient upright walking. Yet the forearm, too, is elongated, as in *A. afarensis* and apes, which means that one Bouri hominid, as yet unidentified, retained an apelike arrangement in the upper limbs. Although these features might seem to be at odds with one another, in fact what they suggest is that the skeleton evolved piecemeal, with the lengthening of the legs preceding the shortening of the forearms.

Arguably even more fascinating than the fossils themselves are the antelope bones bearing cut marks and other signs of butchery that were also found in the vicinity, in addition to a few isolated stone tools. These remains are also 2.5 million years old. But whether *A. garhi* was the hominid who left them behind is unknown, as is the question of whether these ancient butchers were hunters or scavengers. Still, as the oldest cut-marked animal bones on record, they demonstrate that hominids of some ilk were processing carcasses for meat and marrow earlier than previously believed, at a time when hominid brain size was probably still small. This shift in behavior indicates that tool use may have come before brain expansion, and indeed probably permitted that growth by enabling hominids to access the higher-quality foods needed to support a bigger brain. One more aspect of the tools warrants mention. Appropriate raw material for stone tools is rare at Bouri, unlike other sites where cobbles abounded. Perhaps the dearth of artifacts at Bouri is a result of hominids looking after their tools and reusing them, a behavior that archaeologists have called tool curation.

Asfaw and his team contend that *A. garhi* comes from the right place and time to be the ancestor of *Homo*, and that the evidence at the site for stone tool manufacture and butchery—behaviors previously associated exclusively with *Homo*—strengthens that possibility. But that interpretation has not gone unchallenged.

To my mind, it makes sense to view *A. garhi* as a descendant of *A. afarensis*, because in many respects it looks like *A. afarensis*, just with large teeth. If this is true, then we have a lineage of hominids in eastern Africa that persists over 1.7 million years, from *A. anamensis* at 4.2 million years, through *A. afarensis* at Laetoli and Hadar, right up to *A. garhi* at 2.5 million years. But I suspect that rather than giving rise to *Homo*, *A. garhi* may be a side branch on the family tree of humans that evolved large teeth in parallel with the robusts. I think it's unlikely that dentition could have evolved quickly enough to get from the large teeth of *A. garhi,* which lived 2.5 million years ago, to the small teeth of early *Homo,* which emerged roughly 100,000 years later.

Another interesting possibility is that *A. garhi* was an ancestor of the *Homo* species represented by the 1470 skull—the one that Meave thinks *Kenyanthropus* gave rise to. When Richard Leakey declared 1470 to be *Homo,* based on its large cranial capacity of 750 cubic centimeters, there was one dissenter, Alan Walker. He and Meave had spent long hours reassembling the hundreds of fossil fragments that became the 1470 cranium and Alan had intimate knowledge of its anatomy. He drew attention to the massive face and especially the large root sockets in the upper jaw that lack tooth crowns, and argued that 1470 was *Australopithecus.* But Richard's preference for *Homo* won out. Might *A. garhi* have evolved into a hominid like 1470 over the 800,000 years separating them? Perhaps it is time to reexamine Alan's earlier interpretation.

The formative slice of time between 2.5 and 3 million years ago is frustratingly incomplete at the moment. But when more fossil hominids are recovered from that potentially very exciting period I am sure there will be many more surprises. We will finally understand the emergence of *Homo* and, I expect, we will have a better picture of the shape of the human family tree's extinct side branches, where evolutionary experiments in hominid evolution disappeared forever. The difficulty of ascertaining the correct place on the human family tree for a specimen such as *A. garhi* highlights how much more we have to learn.

PART 3

Lucy's Descendants

Ecce Homo

Having spent most of my career studying *Australopithecus,* you might say I have been spoiled. Along with others I have been fortunate enough to recover a sufficient number of informative fossils from the right times and places to be able to reconstruct a remarkably detailed account of the rise and reign of our antecedent genus. We have, by any paleontological standard, an embarrassment of riches. Yet where the early evolutionary history of our own genus is concerned, scientists know surprisingly little. Indeed, the origin of *Homo* is one of the most mysterious chapters of the human odyssey.

We know that our genus emerged between roughly 2 and 3 million years ago, during the late Pliocene epoch, coincident with a major shift in the earth's climate. Before 3 million years ago, Africa was covered with woodlands and forests. But with the ensuing cooling and drying of the continent, grasslands replaced the thick vegetation. As would be expected, the evolution of new kinds of hominids accompanied this environmental change. There were the robust australopithecines, who evolved specializations for subsisting on the tougher plant foods that climate change favored. And then there were the hominids who took the opposite tack, becoming dietary generalists. These omnivorous creatures weren't merely variations on the *Australopithecus* theme, however. These were the founding members of

Homo, a lineage that would diverge so profoundly from *Australopithecus* that it represents what biologists refer to as a grade change. During the late Pliocene, hominids began a journey in which they went from being precultural to being totally dependent on culture for survival in our modern era, and in so doing adopted a whole new way of being proactive about their place in the world. But how did this happen? Although we haven't yet nailed down the details, we know that it had to do with the evolution of that hallmark of humanity: our big brain.

Of all the traits that set our genus, *Homo,* apart from Lucy's, our colossal brain is arguably the most conspicuous. Primates as a group possess large brains and enhanced cognitive capabilities relative to other mammals, with more primitive species like lemurs occupying the low end of the spectrum and chimpanzees and other apes at the high end. But humans took that pattern of braininess to the extreme. Lucy's brain was just shy of 400 cubic centimeters—about the volume of a good-sized grapefruit, and roughly equivalent in size to a chimp's. *H. sapiens,* in contrast, averages a whopping 1,350 cubic centimeters of gray matter. Although the exact relationship between brain size and intelligence is unclear, we can be certain that the more than threefold increase in cranial capacity of modern humans over Lucy translated into radically enhanced brainpower.

Many creatures are bigger, faster, and stronger than humans, but when it comes to wits, we are without peer. These superlative smarts came with strings attached, however. The brain is the hungriest organ in the body, with a metabolic rate sixteen times that of muscle tissue per unit weight. In modern humans, it accounts for 20 to 25 percent of an adult's energy needs. That's more than twice the energy our primate cousins allocate to the brain, and more than four times what other mammals allot.

Considering those hefty energy demands, paleoanthropologists have long sought to understand how the human brain attained such oversize proportions. Thanks to Lucy, we know that it began expanding only after our ancestors began habitually walking upright. But we're not sure what drove the eventual growth. Three explanations have been proffered: a social model, a dietary model, and an environmental model.

The social model starts with the observation that primates, particularly the higher primates, live in societies in which social relationships are very complex. Any individual who is politically successful—from the alpha male of a chimpanzee troop to the president of a corporation—must know every member of his social group. It is vital to recall the history of interactions with those members, to know their personalities, and to be able to predict their future behaviors. A large brain increases the capacity for social learning. Among early hominids, then, those individuals who had larger brains—and thus presumably a keener ability to foster social relationships—left more descendants, according to this hypothesis. Research on another highly social creature, the spotted hyena, seems to support this social intelligence hypothesis. Spotted hyenas live in clans made up of sixty to eighty individuals, each of whom knows the hierarchical position of every animal in the clan. And this species, like humans, has a large, complex brain.

The dietary model, in contrast, focuses on the importance of being able to obtain food. Primates that eat nutritionally rich foods, such as fruits and nuts, have to work hard to obtain these delicacies, because they come in small packets that are generally scattered over the landscape in low densities and are often only seasonally available. These epicureans have larger brains than do primates that subsist on lower-quality foods—leaves, for instance—because reliably locating rarer foods requires the ability to map their distribution. The dietary model assumes that as our ancestors began incorporating more meat into their diet, they faced greater food-procuring challenges, because animals are scarcer than plants—and they run. To consistently obtain meat, so the theory goes, our forebears needed much more gray matter.

For its part, the environmental model moves away from primates as a whole and focuses more narrowly on human evolution, positing that climate change brought about the dramatic increase in brain size. The earliest stone tools and butchered animal bones date to around 2.5 million years ago, when the planet was cooling. According to the environmental hypothesis, the shift in climate transformed the settings in which hominids lived and encouraged our primarily vegetarian ancestors to seek out a new food source—animals—the acquisition and processing of which, as noted above, require a keen intelligence that may well have precipitated our rise to cerebral dominance. It is tempting to view it this way, because within just a few hundred thousand years hominid brain size increased by 50 percent, from 600 cubic centimeters to 900 cubic centimeters.

It remains to be seen which of these theories—or some combination thereof—is the correct one. But given the energy demands of gray matter, it stands to reason that our brain could only expand after our ancestors had adopted a diet sufficiently rich in calories and nutrients. From there, a classic feedback loop was established. The new growth augmented cognitive ability, which allowed for the invention of better tools. Better tools then enabled hominids to obtain more choice food, which in turn permitted the brain to balloon further.

Transformations of other body parts accompanied this brain growth. Indeed, over the course of early *Homo* evolution, the entire hominid skeleton received an extreme makeover. Body size increased, especially in females, reducing the size difference between the sexes in our lineage. And shorter arms and longer legs replaced the long arms and short legs of *A. afarensis*. Also, with easier-to-chew foods on the menu—nuggets of meat and marrow, and perhaps cooked plant foods, such as tubers—mastication requirements eased. This permitted the molars, jaws, and chewing muscles to shrink, morphing the projecting australopithecine face into our flat one. The higher-quality and more easily digested foodstuffs also allowed the gut to dwindle, giving *Homo* a slimmer profile than Lucy's potbellied one. Incidentally, because the gastrointestinal track is metabolically greedy, any reduction in its size would have freed up energy for other demands, such as those of a bigger brain. Therefore, the gut, too, may have had a role in the aforementioned feedback loop.

We know these changes occurred because we can look at the bookends of the transformation—*A. afarensis* and *H. sapiens*—and see the differences between the two. But for a more detailed understanding of this metamorphosis, we must examine the fossil record of the earliest members of our genus, meager though it may be. Until the early 1960s the oldest representative of *Homo* we had to go on was *H. erectus*. Discovered in 1891 by the Dutch anatomist Eugène Dubois along the bank of the Solo River in Java, Indonesia, the type specimen of *H. erectus* (known popularly as Java Man) was the first fossil hominid to be found outside of Europe. The now famous million-year-old fossil—a thick helmet of a skullcap colored a rich, walnut brown by mineralization—turned the spotlight on Asia as the probable birthplace of humanity. Dubois and other European fossil hunters conspicuously avoided Africa in the search for human origins—that humanity could have arisen on the Dark Continent was unthinkable, even though Darwin had proposed just such a scenario twenty years earlier. This Eurocentric view did not envision Africa as a candidate for human beginnings because Africa was considered a backwater populated by "primitive" peoples. Many more fossils of *erectus* have since surfaced at sites across Asia, with specimens ranging in age from 25,000 to nearly 1.8 million years old. But eventually even older examples of *Homo* turned up in Africa.

I distinctly remember the spring day in April 1964 when I learned about the first of these earliest *Homo* fossils. I was a twenty-one-year-old anthropology undergraduate at the University of Illinois at the time. Every week I would go to the library to peruse the latest issue of *Nature* for new paleoanthropological discoveries. This time I was rewarded with an article by Louis Leakey, Phillip Tobias, and John Napier announcing a new member of the human family based on fossils that had been found four years earlier at Olduvai Gorge. I couldn't wait until I got home before perusing my photocopy, so I found a bench bordering the quad where I could sit down and read. "A New Species of the Genus *Homo* from Olduvai Gorge," the report was titled. I glanced at my watch—four in the afternoon here in Champaign-Urbana. But it was nighttime in Africa and I could just imagine Louis Leakey and his wife Mary sleeping at Olduvai, lions prowling around their camp. The name of the place alone seemed incredibly romantic.

I continued reading. Louis reported on two dozen bones of a juvenile male hominid at Olduvai dubbed OH 7. Its large incisors and smallish molars and premolars distinguished it from australopithecines, as did its humanlike hand and foot bones. But what most impressed Louis and his collaborators was their specimen's cranial capacity of an estimated 674 cubic centimeters—roughly 50 percent higher than the australopithecine average—and the fact that stone tools had turned up at the site as well. Prior to the discovery of OH 7, a fossil had to have a cranial capacity of at least 700 cubic centimeters in order to be admitted into the genus *Homo*. But Louis was convinced that his hominid belonged there and, ever the lightning rod, he lowered the cerebral Rubicon to 600 cubic centimeters so that OH 7 could make the cut. He also asserted that this hominid had made the tools at the site, hence the name *Homo habilis,* the "handyman," a name suggested by Raymond Dart. Using culture, in this case the associated Oldowan pebble tools, to define a species was not an accepted practice. For example, early modern *H. sapiens,* often called Cro-Magnon in Europe, is always found with blade tools. But in the Middle East, some early *H. sapiens* have been found with flake tools typical of Neandertals, casting caution on recognizing a species on the basis of associated tools.

I would later learn that Louis's bold claims didn't sit well with a number of paleoanthropologists. Some critics thought it more sensible to broaden the definition of *Australopithecus* to include the Olduvai remains than to stretch the meaning of *Homo;* others considered Louis's find to be simply an older example of *H. erectus,* the only other early *Homo* species then known. Furthermore, another hominid—the robust *Australopithecus boisei*—was also known from Olduvai, and although it made a certain amount of intuitive sense to link the larger-brained *H. habilis* to the tools, there was no way of telling with certainty which of the two manufactured the artifacts based on the available evidence.

Today, more than four decades after Leakey and his collaborators christened *habilis,* many additional fossils have been attributed to the species, and it is a fixture on most renditions of the human family tree. Yet it remains as controversial as ever. Little did I know in 1964 that the *H. habilis* hullabaloo would come full circle for me two decades later. The year was 1986, and the Institute of Human Origins led an expedition to Olduvai at the invitation of the Tanzanian Department of Antiquities. Mary Leakey had recently retired from work at Olduvai and her camp there, which I had stayed at on several occasions in the 1970s, was pretty run-down and needed work, especially the windmills that generated electricity. It felt strange driving into Olduvai with the IHO team, especially since Mary and I had fallen out over the naming and importance of *A. afarensis.*

We had only the most modest expectations for what we might find here in the picked-over hunting grounds of the Leakeys. They had worked there since the 1930s and it seemed wherever I walked I could feel their presence. Just three days in, however, we happened upon a 1.8-million-year-old partial skeleton of a hominid hiding in plain sight, in a spot right next to the dirt road that had taken thousands of scientists and tourists into the gorge. The specimen, OH 62, wasn't pretty: It consisted of more than three hundred fragments, most of which were impossible to reassemble. But we were able to piece together the upper jaw, or maxilla, features of which identified OH 62 as an adult *habilis*. And we had significant portions of the arms and legs. This was especially exciting because it was the first time *H. habilis* limbs and parts of a skull were found in association and we could be certain of the species to which the limb bones belonged. Researchers expected that the limb proportions of *habilis* would presage those of modern humans, whose legs are long relative to their arms. But quite to the contrary, OH 62 had a humerus that was estimated to be 95 percent as long as its femur. By comparison, in a modern human the humerus is 70 percent as long as the femur; in a chimp the two bones are equivalent in length; and Lucy's humerus is 85 percent as long as her femur. Moreover, OH 62 may have only been around a meter tall—a tad shorter, even, than Lucy. In other words, for all of its advanced traits, when it came to body size and limb proportions, *habilis* was downright primitive. Partly as a result of this discovery, and its purported Lucy-like body proportions, experts continue to debate whether *habilis* belongs in *Homo,* or whether it is instead an australopithecine. Reconstructing the limb proportion of OH 62 is tricky because the upper arm bones and thighbones are incomplete. But if this individual had a cranial capacity anywhere close to that of Louis's OH 7, then OH 62 would have had a relatively large brain compared to its body size.

Habilis may have been the first *Homo* species on the evolutionary scene, but it wasn't the only one for long. Within a geological blink of an eye, two more variants of *Homo* had evolved in Africa, *H. rudolfensis* and *H. ergaster.* This is not surprising. Frequently after a breakthrough adaptation—large brains, in this case—there is a branching into several species, what biologists call an adaptive radiation. To borrow an example from elsewhere in the animal kingdom, once birds evolved the ability to fly, they diversified into a multitude of airborne forms. But as is commonly the case in paleoanthropology, scholars disagree over how many species of early *Homo* are represented in the fossil record. Some, for example, maintain the traditional view that *H. ergaster* is simply an African variant of Asia's *H. erectus* and do not recognize *H. ergaster* as a separate species. I have no problem seeing several species of early *Homo* and would not be surprised if paleoanthropologists identify more of them as additional fossils come to light. I believe that since many species within a single genus are differentiated on the basis of subtle differences in anatomy and/or behavior, it is darned difficult to accurately determine the number of species, and I think we will always underestimate the number of species in the fossil record. Naturally we must be cautious not to give a new species name to every fossil hominid specimen found, as has been the case in the past, but we also don't want to move too radically in the other direction of collapsing what are obviously quite different species into one. Somehow we have to strike the right balance between "splitting" and "lumping." The good news is that with our ever-increasing appreciation for diversity within and between species of living and extinct animals, we are better equipped to make assessments of species diversity than ever before. This is why the large collection of *A. afarensis* specimens is so important for determining the number of species in fossil collections.

The species *H. rudolfensis* is based essentially on a single specimen cataloged as KNM-ER 1470, a toothless cranium from a site named Koobi Fora on the eastern shores of Lake Rudolf, now renamed Lake Turkana. This fossil specimen made Louis and Mary's son Richard Leakey a household name in the early 1970s. The 1.8-million-year-old cranium has a cranial capacity of 750 cubic centimeters, well above the *H. habilis* average of 640 cubic centimeters, thus substantiating the notion that 1470 is a member of *Homo*. A few other specimens, including some lower jaws from Koobi Fora and a site in Malawi, are also included in the species. But none of the fossils attributed to *H. rudolfensis* include postcranial bones, so we don't know how large its brain was relative to its body—a critical index when evaluating brain expansion. Richard himself did not assign 1470 to a particular species of *Homo*. It was a Russian anthropologist by the name of Valerii Alexeev who coined the name *H. rudolfensis* based on ER 1470's unique features, such as its tall, very broad face and rather large back teeth (as judged from the tooth roots). The combination of a large brain with more *Australopithecus*-like jaws, face, and teeth seemed to deserve a new species name. But *H. rudolfensis* remains enigmatic and many paleoanthropologists, myself included, have turned to another candidate as a better choice for the ancestor to later *Homo* species.

Although bedeviling in some of their details, *habilis* and *rudolfensis* exhibit the mix of primitive and advanced traits one should expect to see in early *Homo*. The same cannot be said of *H. ergaster*. The Australian biological anthropologist Colin Groves and his Hungarian colleague Vratislav Mazák named this species based on a single mandible from Lake Turkana, although material from Koobi Fora has since been attributed to the taxon as well. This is the second instance in which fossils found by Richard Leakey's team at Lake Turkana were given species names by scholars who had not discovered the actual specimens. But according to the International Code of Zoological Nomenclature, if a new species is properly described and differentiated from other closely related species, then that species name becomes set in stone, in this case *H. ergaster*, "workman."

Despite being only a little younger than the oldest known examples of *H. habilis* and *H. rudolfensis*, some representatives of *H. ergaster* are far more modern-looking than individuals of the other two species. One such specimen—the astonishingly complete 1.6-million-year-old Turkana Boy skeleton from Kenya—reveals an adolescent who had relatively small teeth, a projected adult brain size of 909 cubic centimeters, and long legs, and who would have stood around 1.8 meters tall at maturity. The Turkana Boy demonstrates that by this point in time hominids were well on their way to attaining modern body and brain proportions. Experts generally agree that *H. ergaster* was the ancestor of all later *Homo* species, including our own. But *H. ergaster*'s own origins are veiled in mystery. It is too advanced to be a direct descendant of *H. habilis*. Furthermore, in 2007 a team led by Meave Leakey reported that it had recovered an *H. habilis* jawbone dating to 1.44 million years ago at a site in Kenya called Ileret, east of Lake Turkana, demonstrating that *H. habilis* and *H. ergaster* coexisted in East Africa for nearly half a million years. This makes it additionally unlikely that *H. habilis* begat *H. ergaster*. Where *H. ergaster* came from is, to me, the most intriguing question facing paleoanthropologists today, the answer to which still lies buried somewhere on the African continent.

We have learned quite a bit about ancient *Homo* since Dubois made his famous find along the banks of the Solo River in 1891. Yet the very earliest phases of the evolution of our genus remain largely undocumented. Indeed, as my colleague Bill Kimbel has observed, paleoanthropologists today are not substantially closer to understanding when, where, and under what ecological circumstances *Homo* originated than they were in 1964, when Leakey unveiled *habilis*. The problem is, we have virtually no *Homo* fossils from the time between 2 and 3 million years ago, when the lineage arose. Several putative early *Homo* specimens are claimed to be around 2.5 million years old, including a partial cranium from Sterkfontein, South Africa; a temporal bone from Lake Baringo, Kenya; and a mandible from Uraha, Malawi. Experts disagree over the age of these fossils, however, and whether they truly are early members of our genus. For years, the oldest unequivocal early *Homo* fossil was a 1.9-million-year-old *habilis* cranium from Koobi Fora, Kenya, that Richard Leakey's famed "Hominid Gang" had discovered—a delicately built specimen known as KNM-ER 1813. But in the mid-1990s the Institute of Human Origins unearthed a *Homo* fossil from beyond the 2-million-year mark. Today it is still the only unambiguous specimen from that critical period in time. And it came from, of all places, Hadar.

Ever since the first fossil-hunting expedition to Hadar in 1973, the only hominid we had ever found there was *afarensis*. But in the early 1990s we began to systematically explore the younger sediments of the Hadar Formation, hoping to find fossils that would throw light on the factors that contributed to the emergence of *Homo* and the robust australopithecines. As part of that strategy, we had decided to survey a set of exposures in an area known as the Maka'amitalu, which means "having acacia trees" in the Afar language. We had driven through this area many times since we started working at Hadar, heading for a place we call "the lip," where we descend from the plateau into the Hadar deposits. This part of the trip always made for hair-raising moments because sometimes the Land Cruisers would turn sideways, nearly capsizing. Exiting the deposits was just as trying. There were many times when I would stand watching a vehicle sink deep into the soft soil, wheels spinning and churning up dust, when, in a last-gasp effort, the driver gunned the engine and the car leapt up over the lip, escaping a more dire fate. Whenever we were in Maka'amitalu we were so preoccupied with getting in or out at the lip that we never did any real surveying. We were either rushing in to get camp set up or leaving to head home. That all changed on November 2, 1994.

I wasn't in the field at the time, so as coleader Bill Kimbel was in charge of field operations. The team was getting ready to return to camp after a long day of scouring the area for fossils when two of the Afar collectors, Ali Yussef and Maumin Alehandu, called Bill over to the steep hillock they had been surveying. "Normally when one of the collectors calls, you have reason to believe it's important," Bill remarks. This time was no exception. There on the slope were two halves of a hominid upper jaw, or maxilla, with some teeth still in the sockets. The fossil was very fresh looking, indicating that it had only recently eroded out of the hillside. When Bill fit the two halves together he could tell right away that it wasn't the maxilla of an australopithecine. It lacked the U shape that all australopithecines and chimpanzees possess, in which the dental rows are essentially parallel to each other. And when viewed in profile, their maxilla is thin and projecting, which would have given this hominid a pronounced snout. In contrast, the new Hadar specimen, to which we assigned the catalog number A.L. 666-1, had a parabolic dental arcade, and was much thicker and less projecting in profile. That could mean only one thing. "This was *Homo,*" Bill says, "it was instantly obvious." Less apparent was which species of *Homo* the maxilla came from. He would have to conduct a detailed analysis of the fossil before he could make that determination. The first order of business, however, was to see if he and the others could find any more pieces of the jaw and figure out how old it was. They knew from the stratigraphy that the sediments they were working in were younger than the 3-million-year-old sediments that had yielded the youngest *afarensis* material. But closing in on a more precise date would require some geological detective work.

Bill and the team spent the next two weeks at the site, searching for additional bits of the jaw. Between combing the slope for surface finds and sieving the sediments for pieces that still might be in the ground, they recovered thirty more fragments of jawbone and teeth. "At one point there may have been an entire skull, but as the hill eroded, pieces of the skull were lost over a period of hundreds of years," Bill comments. "What was left behind was the upper jaw." That's not all they found.

Scattered on the surface of the hill were numerous stone tools—simple flakes and cores fashioned from volcanic rock. This was getting interesting. If they could find evidence to link A.L. 666 to the implements, we might have a rare early example of both artifact and maker. The earliest tools on record were the ones from the nearby Gona site that Hélène Roche had begun work at in 1976 and that Sileshi Semaw had subsequently taken over. Those implements were ultimately dated to 2.6 million years ago. No hominid remains were associated with those artifacts, however.

Based on the animal remains found the previous year in the Maka'amitalu basin, we suspected that the maxilla and tools might be significantly older than Richard Leakey's 1813 cranium. But we needed to be certain. We also needed to make sure that these relics came from the same stratigraphic level and that their association on the surface where they were found wasn't merely happenstance. The team made a 2-square-meter excavation into the hill, hoping to spot the exact layer from which the fossil and implements came. They found it: In a layer of silt that matched the silt matrix stuck to the maxilla, workers came upon fourteen implements like the ones that had turned up on the surface. They also recovered three additional nonhuman bone fragments from that level, all exhibiting the same color and patina as the maxilla.

With the geologic source of the jawbone and artifacts identified, we could turn our attention to nailing down the age of the remains. As luck would have it, the hominid- and tool-bearing stratum sat 80 centimeters below a layer of volcanic ash that had already been well dated—the so-called BKT-3 tephra. Using the same argon-argon single-crystal laser-fusion dating technique we employed to pinpoint Lucy's age, geologists had arrived at a very precise age of 2.35 million years for BKT-3. Because the maxilla and stone tools were found below that ash layer, we knew they were at least as old as BKT-3, which was in keeping with the age suggested by the animal species that turned up in Maka'amitalu. The upshot: My team had found nothing less than the earliest well-dated evidence of *Homo* and the oldest ironclad association of hominid remains with stone tools. It was a thrilling development—our finds pushed the fossil record of *Homo* back by nearly a half million years, opening a window on that uncharted time interval between 2 and 3 million years when stone toolmaking and early *Homo* make their first appearance.

Archaeologists have long struggled with trying to nail down the roots of humanity's dependence on technology. Some would argue that tool use must have been present in the common ancestor to apes and humans, since the great apes—most notably chimps— employ a variety of tools, including probing implements for retrieving insects and honey, natural rocks to crack open nuts, and even, in one case, spears for stabbing prey. But intentionally made stone tools fashioned by an early hominid who struck one rock against another to produce a flake don't appear in the geological record until roughly 2.6 million years ago, though I think earlier examples will eventually be recovered. Paleoanthropologists have traditionally attributed stone tools to *Homo* and not australopithecines, but it is entirely plausible that this may change as we dig deeper.

Some of those tools were undoubtedly made of wood or other perishable material and simply did not preserve. Functional studies of the hand bones of *A. afarensis* reveal features such as a long thumb relative to finger length, which would have made this hominid capable of finer manipulation than chimps can manage. *A. afarensis* also had a larger brain relative to their body size than modern apes have. Yet despite years of careful survey at Hadar not a single stone tool has been recovered in association with *A. afarensis*. Furthermore, we have not seen any evidence of cut marks or other modification of animal bone that could have been inflicted by a stone tool. It appears, therefore, that stone tool manufacture and use is a phenomenon that occurred after 3 million years ago.

Based largely on Mary Leakey's remarkable excavations of 1.8-million-year-old archaeological deposits at Olduvai Gorge, the traditional explanation for early stone tool technology is that it is associated with the advent of meat eating. Mary concluded that the association of these Oldowan tools with butchered animal bones reflected a dramatic change in early hominid subsistence behavior. Archaeologists continue to debate whether early hominids, in this case *Homo habilis*, were directly responsible for hunting and killing the animals or might just have been scavenging kills and using flakes to dismember the carcass and other tools to break open the long bones to procure the highly nutritious marrow. There are good arguments for each interpretation. But having spent considerable time in Africa, I tend to favor the idea that hominids were more, rather than less, directly involved with meat procurement. Driving around the landscape in the safety of a four-wheel-drive vehicle I can cover a great deal of territory, and the number of abandoned carcasses I typically see in a day would not keep my companions and me well fed. Furthermore, driving off carnivores such as lions and hyenas from a kill can be pretty scary—life threatening, in fact. To my mind, enhancing one's diet with a regular input of protein-rich meat to feed a hungry hominid brain would have required a more reliable strategy than scavenging.

It is tempting to ascribe all occurrences of stone tools to the processing of meat and marrow, but is it possible that early hominids used these implements for some other activity archaeologists have yet to consider? Here again our closest living relatives, chimpanzees, might offer some insight. Chimps in the Taï forest of West Africa regularly employ unmodified, natural rocks to break open hard-shell nuts. During this activity they generate a scatter of stone flakes that some researchers have suggested mimic an archaeological activity spot. These scatters are very patchy, however, and do not come close to the large tool assemblages that Mary Leakey's excavations at Olduvai revealed. Another distinction is that the stone debris at chimpanzee processing stations consists of what Erella Hovers, the Hadar Research Project archaeologist, calls "angular flakes." Unlike Oldowan artifacts that bear signs of intentional manufacture, such as a striking platform and a bulb of percussion, angular flakes lack these features and appear to be strictly accidental by-products of nutcracking activity. Still, we cannot rule out a eureka moment wherein an early hominid breaking open a hard-cased nut or fruit inadvertently cut himself on a sharp flake, initiating the long road from rudimentary stone tools to modern technology.

Following the exciting discoveries in Maka'amitalu, the next order of business was to see if we could assign A.L. 666 to a particular species of *Homo*. In his analysis, Bill found that the maxilla lacked any characteristics to identify it conclusively as *H. habilis*, *H. rudolfensis*, or *H. ergaster*. But overall it showed the strongest affinity to *habilis*. This was comforting, because *H. habilis* was already believed to be the most ancient species of *Homo*. Additional support for the link to *H. habilis* came in 2003, when Rutgers University paleoanthropologist Robert Blumenschine and his colleagues reported on their discovery in Olduvai Gorge of a 1.8-million-year-old maxilla with a full set of teeth that was assigned to *H. habilis*. The similarities between A.L. 666 and the maxilla from Olduvai (dubbed OH 65) are striking, particularly in the form of the palate and the molars. Still, "whether [666] represents a direct ancestor of *H. habilis* or a lineage broadly ancestral to all *Homo* species, we just don't know enough to say," Bill reflects. Yet even though we can't formally assign A.L. 666 to any named species of *Homo*, we can still deduce a lot about its owner. We know from the shape of the maxilla, for example, that this hominid had a much flatter, more humanlike face than *A. afarensis* did. We also surmise that A.L. 666 was a male, because the maxilla is significantly larger than those of known *habilis* individuals believed to be female, including most of the *H. habilis* specimens from Olduvai. And we know from the wear on his teeth that he was a mature adult when he died. Meanwhile, because the associated artifacts include both flake implements and the cores from which they were struck, we can conclude that our toolmaker passed away in his workshop, as it were.

One more aspect of A.L. 666 bears mention: Its palate is quite wide—a trait that is characteristic of later *Homo* species. "What this is due to no one is quite sure, but it may be related to the expansion of the braincase," Bill surmises, explaining that as the hominid braincase swelled to accommodate ever-larger brains, the adjoining maxilla had to widen accordingly. If so, *Homo*'s trademark braininess may have begun evolving prior to 2.3 million years ago.

Only the retrieval of more fossils as old as or older than A.L. 666 will enable us to classify our cryptic maxilla more specifically—and elucidate the *Australopithecus–Homo* transition. Efforts to do just this are under way, but it's likely to be a tough slog. My team has been looking for more examples of early *Homo* at Hadar, and it will continue to survey the Maka'amitalu. But the younger levels of this area are fairly impoverished as far as fossils go, so we're not holding our breath. And the time period we're really interested in, between 2.5 and 3 million years ago, simply isn't present at Hadar, owing to what geologists term a disconformity—a stretch of time during which sediments either eroded or did not accumulate in a particular place. The only other hominid from this time period is the 2.5-million-year-old *A. garhi* from Bouri in the Middle Awash, which Berhane Asfaw and his colleagues have suggested is descended from *afarensis* and ancestral to *Homo*. But although *A. garhi* is chronologically intermediate between *A. afarensis* and early *Homo*, I doubt that it is a bridge between these two. More likely we are looking at an evolutionary experiment in the hominid family that went extinct. *A. garhi* is known only from Bouri and may have been a local phenomenon that evolved in relative isolation and never got off the ground. There may have been many such failed experiments, as hominids struggled to find their footing in a changing world. If our predecessors did in fact undergo an adaptive radiation prior to 2 million years ago, we're going to have a devil of a time figuring out which species gave rise to *Homo*. And that's assuming we're able to find more hominids from this time period, which at the moment is looking increasingly challenging. But I am hopeful that other parts of the Afar will have geological sediments of the right age and that hominids will be recovered from that vital slice of time.

Leaving the Motherland

W e modern humans are about as cosmopolitan a species as there is. We've established a presence on every continent on Earth, plumbed the ocean depths, and even ventured into space. We are, in fact, consummate explorers. But in stark contrast to us—and, for that matter, to the apes of the Miocene epoch, which roamed throughout the Old World—early humans were homebodies. For most of the roughly 7 million years over which hominids have been evolving, they appear to have lived exclusively on the continent of their birthplace, Africa. Even by Lucy's time, 3.2 million years ago, they were still staying put. Scientists have wondered what it was, exactly, that kept our ancestors cradle-bound for so long. Judging from the dispersal patterns suggested by the fossil record, the occasional disappearance of land bridges due to rising sea levels seemed to periodically restrict the free-ranging Miocene apes. Yet as far as we know, our ancient forebears faced no such obstacles to leaving their natal land. Eventually, though, some brave souls did finally blaze a trail out of Africa, becoming the first in a long line of hominid pioneers.

Until recently, the fossil record had furnished researchers with very few traces of these early wayfarers. With only a smattering of *Homo erectus* fossils from China and Indonesia to go on, we surmised that it was *H. ergaster,* the ancestor of *H. erectus,* who made the first intercontinental forays, just over a million years ago. Tall and large-brained, with long legs and an easy stride, *H. ergaster* certainly looked the part of a pathfinder. In fact, the proportions of this species fore-shadowed those of *H. sapiens.* (In contrast, older hominids—including *H. habilis* and Lucy's kind, *A. afarensis*—were small, short-legged creatures barely brainier than chimpanzees.) But *H. ergaster* emerged in Africa roughly 1.9 million years ago. Why, then, did it take so long for this hominid to make its way off the continent? Researchers thought they had an answer.

H. ergaster, like all tool-wielding hominids before it, manufac-tured implements in the so-called Oldowan tradition. This tried-and-true technology consisted primarily of simple stone flakes that were manufactured with little attention to consistency or detail, but which were surprisingly effective in butchery and breaking up bones to expose their nutritious marrow. Around 1.5 million years ago, however, *H. ergaster* experienced a cultural revolution of sorts and the crudely rendered tools of the Oldowan gave way to the much more specialized and standardized utensils of the Acheulean—including hand axes, picks, and cleavers—which probably provided a better means of butchering. The Acheulean techniques were prim-itive by modern standards, but they were leaps and bounds ahead of those characteristic of the Oldowan. And it was tools made in this advanced manner that showed up at the earliest undisputed hominid locality outside of Africa, a place called 'Ubeidiya in Israel's Jordan Valley that dates to between 1.2 and 1.4 million years ago. Taking that into consideration, our best guess was that humans needed to develop not only long legs and large brains but also cutting-edge technology before they could disperse from Africa.

It was an appealing explanation, and one in which we were rela-tively confident. But a series of spectacular discoveries made over the past two decades at a site called Dmanisi in the former Soviet Republic of Georgia has upended that scenario and left scientists scratching their heads anew over when and why our long-ago fore-bears first set forth from their African fount.

Dmanisi is a world apart from the vast wildernesses of Africa. Tucked in the foothills of the Caucasus Mountains, some 85 kilometers southwest of the Georgian capital of Tbilisi and about 20 kilometers north of the Armenian border, it is a sleepy hamlet, where farmers grow corn and raise pigs, chickens, and goats—the sort of bucolic setting Cezanne might have painted. Amid these modest trappings, however, are vestiges of an erstwhile eminence. In medieval times, the village for which the site is named held a position of tremendous power, thanks to its proximity to Byzantine and Persian trading routes. Today the crumbling tombs of twelfth-century Mongol invaders dot the landscape like so many haystacks. And the stony ruins of a citadel that once stood sentry over the Silk Road commands attention from its perch on a wooded promontory some 80 meters above the confluence of the Masavera and Pineza-uori rivers.

Dmanisi's medieval city has long been the subject of archaeo-logical inquiry. Scholars began excavations there in the 1930s. The first indication that the region might also harbor more ancient relics came to light in 1983, when paleontologist Abesalom Vekua of the Georgian Academy of Sciences happened upon the bones of an extinct rhinoceros in one of the citadel's grain storage pits. The following year, Vekua returned to the site to conduct excavations aimed at retrieving more ancient material. What he found was tanta-lizing: primitive stone tools, but no hominids. It wasn't until 1991 that the team—including a young Georgian paleontologist named David Lordkipanidze—unearthed human remains in the form of an exquisitely preserved mandible, or lower jawbone, with a full set of teeth found under the menacing skeleton of a saber-toothed cat. The size of the mandible and the teeth was reminiscent of *H. ergaster*. Based on the estimated age of the cat and other animals found in the vicinity, the team put the fossil at around 1.6 million years old. If correct, it would be the oldest known hominid outside of Africa.

But when Lordkipanidze and his museum colleague Leo Gabunia brought the specimen to a conference on *H. ergaster* in Germany later that year and showed it to some of the most influential members of the field, they found themselves on the receiving end of serious skepticism. The idea that *H. ergaster* didn't leave Africa until around a million years ago was firmly entrenched, and the mandible from Dmanisi just looked too immaculate to be as old as Lordkipanidze and Gabunia claimed it was. The Georgians had only one option: go back and find more fossils, and build a more persuasive case for their antiquity.

Eight long years passed before more hominid remains surfaced at Dmanisi. But in 1999 the team struck paleoanthropological gold. Workers discovered two skulls in a spot just a few feet from the resting place of the controversial jawbone. The following spring, a paper describing the spectacular finds was published in *Science*. The first specimen—a faceless cranium—appeared to be that of a young adult male; the second, more complete fossil, with its smaller size and more delicate features, seemed likely to be an adolescent or young adult female. Together, the finds revealed a close relationship between the Dmanisi hominids and *H. ergaster*, especially specimens from Lake Turkana, such as the Turkana Boy skeleton. Whereas early hominids from western Europe and eastern Asia possessed regionally distinctive characteristics, the Georgian skulls bore unequivocal resemblances to *H. ergaster* fossils. For example, viewed from above, the bone behind the eyes appears constricted—as though someone pinched the cranium at just that spot—a textbook *H. ergaster* trait.

Meanwhile, the geologists on the team had fortified the dating of the Dmanisi remains. The hominid bones emerged from deposits that lie directly atop a layer of dark gray volcanic rock—the so-called Masavera basalt—that has been radiometrically dated to 1.85 million years ago. The surface of the basalt is relatively unweathered. In fact, it still preserves the flow and cooling features that took shape in the aftermath of the source eruption, indicating that there was little time for erosion to occur before the sediments containing the hominids blanketed the basalt. Under ideal circumstances, fossils are sandwiched between two layers of volcanic material; such rock is relatively easy to date. Dmanisi itself contains only that Masavera basalt. But at another site some 15 kilometers to the west, a second basalt, the Zemo Orozmani, overlies a sedimentary layer that appears to be a continuation of the one containing the hominids at Dmanisi. At 1.76 million years of age, the Zemo Orozmani basalt nicely constrains the date range for the fossils. Paleomagnetic analyses of the fossil-bearing sediments themselves yielded compatible results, concluding that they were deposited nearly 1.77 million years ago, when the planet's magnetic polarity underwent a reversal—an event known as the Olduvai-Matayuma boundary. And the menagerie of creatures associated with the hominids at Dmanisi—species of ostrich, gazelle, jaguar, musk ox, elephant, giraffe, and hyena among them—are known to have lived between 1.6 and 2.0 million years ago.

The additional fossils and the robust dating results erased any doubts about Dmanisi taking pride of place as the oldest undisputed hominid site outside of Africa, and in so doing pushed the colonization of Eurasia back almost half a million years. They also sounded the death knell for the theory that a shift to the more advanced Acheulean technology is what precipitated the exodus. The excavators had recovered more than one thousand stone artifacts from nearly the same stratigraphic level that yielded the mandible in 1991, and there wasn't a single Acheulean implement in the bunch. Rather, the collection comprises the simple flakes, choppers, and scrapers—all made from the same local basalt—that typify Oldowan tool assemblages in East Africa.

It was a defining moment for Lordkipanidze. The son of a

classical archaeologist, he began working at Dmanisi in 1991, having returned to his native Georgia after completing his Ph.D. in Russia. He was just twenty-eight years old when he assumed leadership of the excavation and helped unearth the mandible that put Georgia on the map. Today the tall, ruddy-cheeked Lordkipanidze is also director of the Georgian National Museum in Tbilisi.

Lordkipanidze's considerable efforts and those of the international team he heads have continued to pay off. In July 2002 the fossil hunters reported that they had recovered another virtually whole skull, including the mandible. It is one of the most primitive *Homo* skulls ever discovered outside of Africa. The first two Dmanisi skulls had cranial capacities of 770 cubic centimeters and 650 cubic centimeters. The third one, however, had room for just 600 cubic centimeters of gray matter. At less than half the size of a modern human brain, this is teeny even by *H. ergaster* standards. And its small braincase isn't the only primitive feature. The third skull exhibits a more delicate brow and snoutier midface than is typical of *H. ergaster*. Those features—along with the curvature of the rear of the skull—are more reminiscent of *H. habilis*, the putative ancestor of *H. ergaster*.

With the discovery of the third skull at Dmanisi, a second pillar of the textbook theory of what finally spurred hominids to start globe-trotting gave way. Hefty brains were not, in fact, required for intercontinental migration. Some of these colonists were barely brainier than *H. habilis*. And yet they appear to have been quite humanlike in some aspects of their behavior, as another find would illustrate to dramatic effect.

In 2005, Lordkipanidze's team unveiled a fourth skull from Dmanisi—complete with a matching mandible. Like most of the Dmanisi material, this specimen was in superb condition, showing none of the signs of weathering that so often afflict fossils of such antiquity. Yet it was missing all but one of its teeth—the lower left canine. Through careful study of the jawbones and tooth sockets,

the team determined that the individual—believed to be male—
had lost his teeth several years prior to death, as a result of aging or
disease or a combination of the two. This specimen is the oldest
known example of a hominid whose ability to chew was so severely
compromised. In fact, no other hominid on record exhibits such
profound tooth loss and the attendant remodeling of the jawbone.
How could this poor fellow have survived for so long without teeth?
His remains turned up near a bunch of stone artifacts and bones
bearing cut marks and percussion marks associated with butchery
and meat eating at other early hominid sites. Lordkipanidze and his
collaborators thus propose that he may have favored soft plant and
animal foods, possibly including bone marrow or brain. But they
also raise another provocative possibility: He may have relied on
other members of his group to provision him with easy-to-eat foodstuffs.

This wouldn't be the only example of compassion among early
hominids. Bruce Latimer of the Cleveland Museum of Natural History and others have argued that the Turkana Boy suffered from a
condition that stunted the growth of the openings in his vertebrae,
the so-called vertebral canals. The resulting compression of his
spinal cord would have made it difficult for him to get around. That
the Turkana Boy survived past childhood may indicate that members
of his social group protected and provided for him.

H. ergaster generally—and the Dmanisi people specifically—
may have been cognitively humanlike in other ways, too. Received
wisdom holds that *H. ergaster* was incapable of language. Its vertebral
canals were simply too small to accommodate a spinal cord large
enough to permit sufficient control over the muscles involved in res-
piration and vocalization. It turns out, however, that argument was
based on an anomalous specimen, none other than the Turkana
Boy. His vertebral canals are no larger than a chimp's. But when
Marc Meyer of Chaffey College, along with Lordkipanidze and
Vekua, studied the vertebrae that turned up at Dmanisi, they found

their canals to be proportioned like a modern human's. If the small size of the Turkana Boy's vertebral canals is a result of a pathological condition, then the large size of the Dmanisi examples is a better yardstick for assessing *H. ergaster*'s ability to speak. The Dmanisi hominids, Meyer says, had spinal cords identical in size and shape to those of modern humans that would have permitted the control of the diaphragm, intercostal muscles, and abdominal muscles necessary for speech. This doesn't mean that *H. ergaster* definitely could converse, because language depends as much, if not more, on cognition as it does on the physical capacity for speech. But it does reveal that there were no neurological limitations on this hominid's ability to talk. And it raises the possibility that the gift of gab helped these early explorers push into new territories.

The small brains and modest tools of the Dmanisi people startled paleoanthropologists. But there were even more revelations to come. Experts were eagerly anticipating description of postcranial bones that had been recovered at the site. Would they have the small body size and primitive limb proportions of *H. habilis* and the australopithecines, or would their measurements be closer to those of our modern frame? Given the modern body size and relative limb lengths of the most complete *H. ergaster* individual on record, the Turkana Boy, researchers posited that humans needed to evolve large bodies and long legs before they could leave the African continent. The Georgian fossils could make or break this widely held presupposition.

In September 2007 Lordkipanidze and his colleagues published a paper in *Nature* describing the postcranial remains from Dmanisi. The team had retrieved a partial skeleton believed to be that of the same adolescent represented by the skull with the 600-cubic-centimeter cranial capacity that they had unveiled in 2002. They also recovered skeletal elements of three adults. Together, the bones reveal a pastiche of primitive and modern traits. On the archaic side, the Dmanisi people appear to have been fairly petite, standing just 145 to 166 centimeters tall and weighing an estimated 40 to 50 kilograms. That's a lot larger than Lucy, but diminutive for a modern human. Also primitive is the upper arm bone, or humerus, which is straight instead of twisted like ours, revealing that these hominids would not have been as good at throwing as we are. And a fragment of right shoulder blade resembles that of the great apes in a number of features, including a shoulder socket that faces upward as in australopithecines rather than sideways as in *Homo*. In these ways, the upper limb of the Georgian hominids is rather like that of the australopithecines, which is to say it retains a number of adaptations to life in the trees. But these arboreal traits are merely evolutionary baggage handed down from a more primitive ancestor—there's no reason to think that these efficient bipeds spent time in the trees. Likewise, their brain size relative to their body size—a measure known as the encephalization quotient—is comparable to that of *H. habilis* and the australopithecines. On the other hand, the Dmanisi folks had limb proportions resembling our own, with long thighbones relative to their upper arm bones. This indicates that these hominids had more energy-efficient bipedal locomotion than that of the australopithecines because the energy cost is inversely proportional to the length of the lower limb in bipeds. And their spine was more similar to that of modern humans than to that of australopithecines, suggesting that it was adapted to the increased compression loads that come with long-distance walking and even running.

Although a number of the earlier ideas about what it was that finally permitted our predecessors to leave Africa and begin peopling the rest of the world have wilted in the face of the discoveries at Dmanisi, one is still standing—and has in fact gained credence from the analyses of the skeletal remains. Whereas the primitive characteristics of the Dmanisi hominids are likely holdovers from *H. habilis* and the australopithecines, the modern traits—including longer legs—seem to be the result of selection for more efficient long-distance walking. On current evidence, then, it seems that evolved limb proportions were a prerequisite for intercontinental peregrinations. If in the future fossil hunters discover short-legged hominids outside of Africa, however, we'll have to go back to the drawing board on this issue. (Another viable explanation for how hominids finally succeeded in spreading north of Africa is that a shift from the largely vegetarian regimen of Lucy and her brethren to a hunter-gatherer diet permitted hominids to persevere in higher latitudes over the winter months, when plant foods are in short supply.)

In the meantime, there are more Dmanisi mysteries to unravel, including the question of how, exactly, these ancient Georgians are related to other hominids. To Lordkipanidze, the mosaic of primitive and advanced traits in the fossils suggests that they are the progenitors of *H. ergaster*—either by virtue of being among the very earliest members of this species or by representing a new species, *H. georgicus*, that was directly ancestral to it. One could call it a sort of missing link between *H. ergaster* and *H. habilis*. There are other interpretations of the Georgian finds, however. Considering the anatomical variation evident in the skulls and jawbones, Jeffrey Schwartz of the University of Pittsburgh has suggested that more than one hominid species may have called Dmanisi home. By way of example, he pointed to a massive mandible, unearthed in 2000, that is far larger than the others that have turned up. In Schwartz's view, a single species/population of *Homo* cannot comfortably encompass both an individual this huge and one as dainty as the one represented by the skull with the 600-cubic-centimeter cranial capacity. But Lordkipanidze and longtime Dmanisi team member Philip Rightmire of Harvard University countered that a much more likely explanation for the disparity is that *H. ergaster* was just far more variable than previously believed. After all, the fossils all hail from the same stratigraphic layer at the site. It may well be that the small individuals from Dmanisi are all females and the enormous mandible belonged to a male. If so, the differences between the sexes in *H. ergaster* may have been significantly more pronounced than previously thought—not unlike that between Lucy and her male peers, in fact.

Conventional paleoanthropological wisdom holds that the emergence of *H. ergaster* marked a radical departure from previous hominid body plans and the first instance of a hominid that was, in many ways, like us. In addition to being tall, large-bodied, and long-legged, *H. ergaster* was also believed to resemble humans in having relatively few differences between males and females, beyond the obvious sex characteristics. Size is one such differentiator. Whereas the australopithecines exhibit marked sexual dimorphism—comparable to that of gorillas, in which males are nearly twice the size of females—*H. ergaster* males were thought to be only around 10 to 20 percent larger than the females. Yet over the past few years, fossil hunters have discovered several surprisingly small *H. ergaster* specimens, not only in Georgia but also in Kenya. In 2004 Rick Potts of the Smithsonian and his colleagues reported that they had found a 900,000-year-old partial skull of an *H. ergaster*–like hominid with an estimated cranial capacity of less than 800 cubic centimeters at a site called Olorgesailie in Kenya. And in 2007 Meave Leakey and her collaborators announced that they had collected a 1.55-million-year-old partial cranium from a site east of Lake Turkana called Ileret that, in its external dimensions, is the smallest *H. ergaster* on record. (It does not have the smallest cranial capacity, however; that distinction still belongs to the Dmanisi skull with the 600-cubic-centimeter cranial capacity.)

Yet even if the Dmanisi males were quite a bit larger than the females, as the giantic jawbone suggests, these people were still significantly smaller on average than *H. ergaster*. And this fact poses further problems for paleoanthropologists trying to figure out not only how but also *why* hominids left Africa in the first place. One popular theory holds that the large body size of *H. ergaster* required that this hominid adopt a higher-quality diet than that of its petite predecessors in order to meet its increased energy needs. Specifically, this bigger hominid would have had to eat meat. And because carnivorous animals must search farther and wider for food than vegetarians, so too would *H. ergaster* have needed to expand its horizon. This pursuit of prey, so the hypothesis goes, may have been what led *H. ergaster* into Eurasia. Needless to say, the discovery of small-bodied hominids outside of Africa spells trouble for this scenario. But maybe *H. ergaster* simply evolved shorter proportions when it arrived in Georgia, which is and was much colder than the Turkana region. That, after all, is what the cold-adapted Neandertals did.

Whatever it was that sparked wanderlust in these hominids, it's easy to see why they put down roots in southern Georgia. Flanked by the Black Sea to the west and the Caspian Sea to the east, the region would have enjoyed a mild climate akin to that of the Mediterranean. Reconstructions of the environment at Dmanisi as it was 1.77 million years ago reveal an area that was tremendously diverse, with fossils of woodland animals such as deer and grassland beasts such as horses, indicating a mix of forest and savanna habitats. And the presence of two rivers at the site and a lake nearby meant a reliable water supply—both for the hominids and potential quarry. Cut marks evident on animal bones from Dmanisi make clear that the hominids ate meat at least occasionally. But whether they obtained the meat by hunting or by scavenging it from kills made by the many large carnivores that lived in the area—cats, hyenas, and wolves among them—is not yet known.

Abundant though the natural resources were at Dmanisi, life wasn't entirely carefree for its prehistoric human residents. In addition to facing stiff competition from the local carnivores for access to meat, the hominids sometimes ended up as dinner themselves, as feline gnaw marks on the giant mandible and puncture wounds on one of the skulls attest. In fact, most of the hominid remains have turned up in a particular section of the site (dubbed the champagne room, for the many celebration-worthy finds made there), leading the team to speculate that denning saber-toothed cats may have amassed them.

The remarkable finds from Dmanisi establish that hominids made it out of Africa by at least 1.77 million years ago. What we don't know is where they went from there. There are hints of a similarly ancient hominid presence in Indonesia, from a site called Modjokerto in Java, where the partial skull of a five- or six-year-old child turned up in 1936. The most recent age estimates, based on the argon-argon dating method, place the fossil at 1.8 million years old. And other *H. erectus* fossils from the site of Sangiran, some 240 kilometers away, have been dated to 1.6 million years ago using the same technique. But some experts have questioned whether the sediments that were dated hail from the same locations as these fossils, arguing that the bones may be much younger than the argon dates would suggest. Meanwhile, the earliest European fossils—announced in March 2008—are 1.1 to 1.2 million years old and come from a cave called Sima del Elefante located in northern Spain's Atapuerca Mountains. Considering their physical characteristics, the Dmanisi colonists could conceivably be ancestral to later *H. erectus* from Asia. (In fact, there are Oldowan-like stone tools in China's Nihewan Basin that, according to paleomagnetic evidence, are more than a million years old. Their age is uncertain, however, owing to a lack of material suitable for radiometric dating.) But Georgia might just as easily have been the last stop on their grand tour, making this particular group an evolutionary dead end.

It would be vastly oversimplifying matters to assume that our ancestors populated the rest of the Old World in a single wave of migration. There were undoubtedly many movements out of Africa—as well as some back in. In fact, one theory that has been gaining visibility in recent years submits that our genus actually originated in

Asia, not Africa. To be sure, this is an antiestablishment proposition. That *Homo* arose in East Africa is one of paleoanthropology's most fundamental precepts. But in a paper published in *Nature* in 2005, Robin Dennell of the University of Sheffield in the United Kingdom and Wil Roebroeks of Leiden University in the Netherlands threw down the gauntlet. They pointed out that not a single early Pleistocene hominid fossil or stone tool has turned up in the vicinity of either of the two putative migration routes out of Africa into Asia: along the Nile Valley or across the southern end of the Red Sea. Yet the Dmanisi finds show that surprisingly primitive hominids were living in southwest Asia by 1.7 million years ago. If large bodies and large brains weren't a prerequisite for intercontinental traveling, and if the australopithecines had adapted to life on the African savannah, then it stands to reason that they could have spread into the comparable Asian grasslands as early as 3.5 million years ago. Just because no one has found hominids older than *H. ergaster* outside of Africa doesn't mean they're not there, Dennell and Roebroeks argue. Or, to echo a common saying among fossil hunters, absence of evidence isn't evidence of absence. And although that might sound like special pleading, the fact is that, as Dennell and Roebroeks observe, paleoanthropologists have explored Asia far less intensively than they have Africa.

So far, however, African *Homo* predates Asian *Homo* by a long shot—a good half million years, to be precise. Thus, to my mind, Africa is the homeland for *Homo* and indeed *H. erectus*, or as it is called in Africa, *H. ergaster*. The idea that *H. ergaster* originated in Asia and then migrated into Africa is at this time speculation because no hominid more ancient than *H. erectus* has been found in Asia. The Georgian hominids bear an undeniable resemblance to *H. ergaster* in eastern Africa, and until concrete evidence is found in Asia, it is more reasonable to think that *H. ergaster* evolved in Africa.

Future discoveries at Dmanisi should cast light on this issue. And no one doubts that more stunning finds will come from this place, which is now widely recognized as one of the most important hominid localities in the world. Thus far workers have combed through only a tiny fraction of the site, yet they have already unearthed more spectacular hominid fossils than anyone could ever hope to find in a lifetime. Lordkipanidze has his hands enviably full. In 2005 I had the pleasure of meeting him at a conference in Rome on early *Homo* in Europe. He concluded his presentation by noting that it was my first book on Lucy (translated into Russian) that inspired him to become a paleoanthropologist. I can only imagine how influential his own finds have been on the next generation of fossil hunters. I'll be watching eagerly to see what other treasures his team uncovers.

The Hobbits of Flores

Every once in a while, there comes to light a fossil that shakes the foundation of paleoanthropology to its very core and forces us to reconsider what we thought we knew about human evolution. Lucy was one such discovery. Her unprecedented antiquity, preservation, and primitiveness came as nothing short of a revelation. Lucky for us, there have been many spectacular finds since Lucy. But perhaps none is quite so remarkable as the recovery just a few years ago of an enigmatic partial skeleton nicknamed the hobbit, after the fanciful characters of J. R. R. Tolkien's novels. The hobbit lived as recently as 17,000 years ago—just yesterday, in geological terms— yet the specimen is so unlike us that its discoverers have considered assigning it to an entirely new genus of hominid. Indeed, the strange skeleton exhibits striking similarities to, of all fossils, the eons-older Lucy herself.

The story of the hobbit begins on the tropical island of Flores, located some 200 nautical miles east of Bali in the Indonesian archipelago. Ever since the Dutch anatomist Eugène Dubois discovered in 1891 the remains known as Java Man, researchers have toiled to wrest more ancient bones from Indonesian ground. Michael Morwood, a silver-bearded archaeologist at the University of New England in Armidale, New South Wales, Australia, started working there in 1995, excavating a site in the Soa Basin of central Flores called Mata Menge that contains 800,000-year-old stone tools believed to be the handiwork of *Homo erectus*. The discovery was startling, because it suggested that *H. erectus* had somehow managed to cross the deep waters separating Flores from Java. To Morwood, who has spent his career studying the evolution and migration of humans through the Indonesian archipelago for insight into the peopling of Australia, this could mean only one thing: *H. erectus* built watercraft. The notion of *H. erectus* as seafarer came as a shock: Previously it was thought that humans only started building boats 40,000 to 60,000 years ago, which enabled our ancestors to colonize Australia. But little did Morwood know that Flores had an even bigger surprise in store.

Morwood eventually came to explore a large limestone cave in the western part of Flores called Liang Bua, "cool cave" in the local Manggarai language. This was not the first time archaeologists had worked in Liang Bua. In the 1970s, Raden Pandji Soejono of the Indonesian Center for Archaeology in Jakarta discovered artifacts dating to the Mesolithic, the cultural period that spans the time from roughly 8500 to 4000 B.C. But the archaeological deposits clearly extended well beyond those Mesolithic levels and no one had tried to reach bedrock. Morwood and Soejono resolved to plumb Liang Bua's depths and, with a team of Australian and Indonesian scientists, resumed excavation of the cave in July 2001—a decision that would prove more fruitful than any of them could possibly imagine.

In July 2002 the team returned to Liang Bua for a second field season. This time, in levels spanning the time from 60,000 to 12,000 years ago, the relic hunters hit upon the bones of rats, birds, an extinct elephant relative known as *Stegodon,* and a bounty of stone tools far more primitive than those from the overlying Mesolithic levels of the cave. Diagnostic remains of the people who made these tools and butchered these creatures proved more elusive, however.

The previous year, excavators had recovered a radius—the smaller of the two bones in the lower arm—at a depth of 6 meters. Smaller and more curved than your average modern human radius, the bone was intriguing, but it lacked any telltale features to link it to a particular hominid species. It wasn't until the tail end of the third season at Liang Bua that the team recovered any remains that might be able to reveal the identity of the pre-Mesolithic inhabitants of the cave. The first bit was modest—a single premolar found amid a dense deposit of stone artifacts and *Stegodon* bones. Scientists can sometimes recognize species on the basis of a single tooth, thanks to the distinctive features of mammalian dentition. But none of the team members on-site had sufficient expertise in anatomy to assign the premolar to a particular hominid; Morwood would have to seek outside counsel for that assessment. He didn't have to wait long: Though there was still one week of fieldwork left in the season, Morwood was due to return to Java to take care of paperwork. From there he would fly home to Australia—a copy of the mystery tooth in tow.

Meanwhile, back at Liang Bua, the Indonesians, working under the supervision of Thomas Sutikna of the Indonesian Center for Archaeology, were wrapping up the field season in his absence. Morwood would soon have far more than a tooth to study. Three days after he left the site, one of the Manggarai workers on the team found the better part of a hominid skeleton in sector 7 of the cave, embedded in a layer of sticky brown clay. The bones were extremely fragile—the consistency of wet tissue paper. As such, the archaeologists would have to exercise great care in removing them from the surrounding clay, hardening them with a mixture of glue and acetone as they went. Based on the small size of the skeleton and the primitive appearance of the skull, with its sloping forehead and thick cranial bones, the Indonesians believed the remains to be those of a nonmodern child. Could this be the hominid that fashioned the tools and hunted the *Stegodon* and other creatures?

Morwood brought a cast of the tooth and news of the skeleton back to Australia with him, to share with his colleague Peter Brown, also at the University of New England. Brown's assessment of the tooth: broadly humanlike but not from a modern human. He was skeptical about the skeleton, based on the scant details they were able to obtain over the phone. A specimen of this age seemed almost certain to be a modern human. But the tooth—with its two roots, instead of the usual one, along with other primitive features—did suggest something interesting was going on in Liang Bua. Intrigued, he joined Morwood in boarding the next plane to Jakarta.

Brown, a respected anatomist, led the analysis of the skull and partial skeleton, known as LB1. Judging from the pelvis anatomy, LB1 was a female. And given her stage of dental development, she was almost certainly an adult when she died. But in nearly all other respects, the specimen baffled. It had a braincase the size of a chimp's and shared Lucy's broad, flared pelvis. And she stood hardly more than a meter tall. Yet other characteristics including the small teeth, narrow nose, and the long, low, broad skull called to mind *H. erectus*. LB1 also resembled archaic hominids—and differed from modern humans—in lacking a chin.

Brown pored over the bones for three months before reaching a verdict. Then, in October 2004, he and Morwood and their colleagues published their findings in *Nature*. Their conclusion was stunning: LB1—as well as the isolated tooth and the arm bone found in deeper deposits in Liang Bua—represented a new species of human, *Homo floresiensis*, that lived alongside our own species for millennia. The discovery of a new hominid species is always exciting, but finding one that overlapped in time with *H. sapiens* is much more thought provoking, for it raises the question of how the two groups interacted, if at all. Previously, scientists thought that *H. sapiens* had been comfortably ensconced in the role of sole hominid species by that time, following the demise of the Neandertals in Europe and *H. erectus* in Asia some 25,000 years ago. The Flores bones, however, indicated that *H. sapiens* shared the planet with another species of human from around 95,000 years ago to 18,000 years ago.

Even more startling than the idea that *H. floresiensis* lived such a short time ago was the fact of how different the creature apparently was from us. *Homo erectus* and Neandertals—the only two species *H. sapiens* was believed to have overlapped with prior to the discovery of the Liang Bua bones—were similar to our modern human forebears in having relatively large bodies and large brains. Neandertals stood around five and a half feet tall and had an average brain size of 1,400 cubic centimeters. *Homo erectus,* for its part, had a mean stature of five foot ten, and a brain size of 900 cubic centimeters. LB1, however—with a height under three and a half feet and a brain size of around 400 cubic centimeters—was proportioned much more like little Lucy and other australopithecines. How could such a Lilliputian hominid have arisen so recently in human evolution?

Brown and his collaborators had an answer. In their paper, they contended that *H. floresiensis* was a descendant of *H. erectus* that had become stranded on Flores, where it evolved in isolation into a sort of miniature *erectus*. This is not as improbable as it might sound. Biologists have long known that mammals larger than rabbits have a tendency to shrink on small islands, probably as an adaptation to the limited availability of food in these environments. Dwarfing poses few risks to these animals, because predators are rare in such cozy locales. Mammals smaller than rabbits, in contrast, tend to enlarge under such conditions. Flores itself bears witness to several examples of this so-called island rule: Researchers have found dwarf representatives of *Stegodon*, as well as remains of rats the size of rabbits. But the possibility that humans might be subject to the same "island rule" flew in the face of a long-cherished notion among paleoanthropologists; namely, that culture has largely insulated us against the selective pressures that shape other animals. Case in point: Whereas most mammals have evolved thick fur in response to cold conditions, humans have built fires and made clothing.

What makes the hobbit all the more perplexing is that in addition to having such a petite body, she has a tiny brain. Over the course of human evolution, brain size has trended steadily upward, from 360 cubic centimeters in the earliest putative hominid (*Sahelanthropus*) to 1,350 cubic centimeters on average in modern humans. This makes sense because gray matter is the primary hominid adaptation. Even modern pygmies with their small bodies have brains that are the same size as those of modern humans of average stature. In modern humans, the onset of puberty marks the end of brain growth and a final spurt in body growth. Pygmies have normal growth patterns until puberty, after which their bodies stop growing. But because brain growth in our species is completed by then, it is only their bodies—not their brains—that are stunted. This is why pygmies have heads that appear oversize in relation to their bodies. The hobbit's brain, however, is far smaller than one would expect for a dwarfed descendant of *H. erectus*. In fact, according to primatologist Robert Martin of the Field Museum in Chicago, for LB1 to be a miniaturized form of *H. erectus*, it would have to have been just one foot tall with a body weight of only four pounds to explain such a diminutive brain.

To complicate matters further, the hobbit remains turned up in association with stone tools. Most are simple stone flakes struck from volcanic rock and flint—akin to implements made by early *Homo*. But some are quite sophisticated—comparable, even, to implements manufactured by early members of our own species. Morwood and his colleagues described the archaeological and paleontological remains from Liang Bua in a separate paper in *Nature* that accompanied Brown et al.'s report. In it they revealed that the hobbits appear to have manufactured points and blades, some of which Morwood believes were employed as spearheads and knives for hunting and butchering. The team also found awls, which the mini hominids could have used to make holes in wood or hides. These advanced Liang Bua tools only showed up in association with pygmy *Stegodon* remains, suggesting that hunting *Stegodon* was an important part of hobbit survival. Cut marks on bones belonging to Komodo dragons, rats, bats, and birds indicate that they dined on other species as well. They also seem to have mastered fire: Excavators found signs of ancient hearths and scorched animal bones in the same levels that yielded the hominids.

Taken at face value, the finds indicate that despite having tiny brains, the hobbits were clever enough to craft fancy tools, cooperatively hunt large, dangerous animals (even a pygmy *Stegodon* would have been giant quarry for hobbits), and butcher and roast their kills. And intriguingly, more recent work by Morwood's group has found striking similarities between the advanced tools at Liang Bua and ones from much older sites, like 800,000-year-old Mata Menge. These sites, too, contain remains of *Stegodon,* albeit the large-bodied variety. Morwood believes that the hobbits may be the genetic and technological heirs of the hominids that left behind the remnants at these much more ancient sites.

The announcement of the stranger-than-fiction discovery on Flores sparked a media frenzy, with hobbit headlines making the front pages of newspapers and covers of magazines around the globe. But controversy nipped at the heels of Morwood and Brown's bold claims. Within days of the *Nature* publication, Australia's *Sunday Mail* ran a letter from paleoanthropologist Maciej Henneberg of the University of Adelaide claiming that LB1 was merely a modern human that suffered from a condition known as microcephaly (from the Greek for "small brain"). Furthermore, he argued, the isolated arm bone found deeper in the cave corresponds to a height of between 1.51 and 1.62 meters—within the range of modern humans—showing that taller individuals, too, lived at Liang Bua. Henneberg's was not the only voice of dissent. Alan Thorne of the Australian National University was similarly dismissive, as was Indonesia's reigning paleoanthropologist, the then-seventy-five-year-old Teuku Jacob of Gadjah Mada University in Yogyakarta (Jacob has since passed away). Skeptics were to be found in the United States, too, notably Robert Martin of Chicago's Field Museum and Robert Eckhardt of Pennsylvania State University. All of them believed LB1 was not a new hominid species at all, but rather a modern human with a pathology.

What happened next in the saga of the hobbits vividly illustrates the occasional perils of working in this field, in which the stakes are high and the egos are fragile. Through a series of events—the details of which are disputed—Jacob had the delicate bones transported from their repository at the Indonesian Center for Archaeology in Jakarta to his laboratory, against the wishes of members of the discovery team. Morwood and company contend that he effectively kidnapped the remains; Jacob and his compatriots insisted that they did nothing wrong and had the full cooperation of Morwood's teammate, Soejono, a longtime associate of Jacob's. While in possession of the remains, Jacob invited several colleagues to study them with him, including Henneberg and Eckhardt. He also furnished Jean-Jacques Hublin of the Max Planck Institute for Evolutionary Anthropology in Leipzig with two samples of bone for possible DNA extraction and analysis. Jacob's sharing of the material with his colleagues further enraged Morwood, who felt that the discovery team should have priority access to the finds.

·   Although some researchers doubted that Jacob—who kept virtually all of Indonesia's most precious hominid specimens in a vault in his lab and had a reputation for not sharing—would ever return the remains to Jakarta, he eventually did give them back, albeit well beyond the deadline that was set for their return. But when the bones arrived at the Indonesian Center for Archaeology, many were seriously damaged. A newly recovered lower jaw was broken and sloppily reassembled, and LB1's pelvis—previously in relatively good condition—was shattered. Morwood's team promptly accused Jacob of mishandling the precious remains. Jacob's group, however, maintained that the bones suffered injuries en route to Jakarta, not in his lab. The situation became so vicious that the Indonesian government placed a moratorium on further excavation at Liang Bua until the two sides cooled off.

A report detailing one team of skeptics' conclusions appeared in the September 5, 2006, *Proceedings of the National Academy of Sciences.* In it Jacob, Eckhardt, and their collaborators argued that some of the characteristics said to be unique to LB1 (and therefore evidence that she represents a species new to science) are found in modern Australomelanesians. These traits include the absence of a chin. Moreover, they contended, LB1 bears numerous signs of abnormal development—including the diminutive braincase, asymmetries in the skull and skeletal bones, and weakly marked sites for muscle attachment on the arm and leg bones. They concluded that LB1 was a pygmy modern human with a form of microcephaly. Furthermore, they claimed, some of LB1's other allegedly primitive features can still be found in people who live on Flores today. In fact, according to these hobbit deniers, a group of pygmies who live in a village called Rampasassa, located near Liang Bua, are probably the modern-day descendants of the population to which LB1 belonged.

Peter Brown counters that all modern humans—including microcephalics and pygmies—have chins. And the so-called skull asymmetries are the result of distortion that occurred while the skeleton was buried under 9 meters of sediment. Even so, a sophisticated analysis of LB1's skull shape conducted by Karen Baab of Stony Brook found that LB1's skull is not unusually asymmetrical, especially when compared to other fossil *Homo* skulls. As for the asymmetries allegedly visible in cross section in the leg bones, "[Eckhardt] is dead wrong," retorts William Jungers, also at Stony Brook, who has studied the remains from Liang Bua extensively. "Asymmetry in the femora and the tibiae are modest at best, and well within the range encountered by normal modern humans." Upper-limb expert Susan Larson, another member of the Stony Brook group, is likewise unpersuaded by the pathology argument. There is no simple relationship between the size of the muscle markings and the strength of the muscles, she explains.

Not to be deterred, the hobbit defenders have conducted further studies of their own on the bones and buttressed their assertion that the remains represent a hominid species new to science. Much of that research has focused on the question of microcephaly. Although investigators can't study LB1's brain directly, they can examine it indirectly via a cast of the impression her brain left on the braincase. Such endocasts, as they're known, are typically obtained by making a latex mold of the interior of the braincase and then creating a plaster cast that replicates the morphology of the brain. LB1's skull is too fragile for this procedure, so Brown created a virtual endocast using CT scans of the braincase, and then rendered the endocast in resin using a rapid-prototyping technology known as stereolithography.

Dean Falk of Florida State University, an expert on hominid brain evolution, led the analysis, comparing LB1's endocast with those of great apes, *H. erectus,* australopithecines, modern humans of average stature, and pygmy and microcephalic modern humans. In a paper published in *Science* in March 2005, her team reported that their findings bolstered the hypothesis that LB1 does, in fact, represent a novel hominid species, not a diseased modern human. The researchers determined that in terms of brain size relative to body size, LB1 was most like an australopithecine. But looking at the shape of her brain, *H. erectus* is the closest match. Falk, who expected LB1's endocast to look like a chimp's based on the small size of the organ, was shocked to instead find advanced features like expanded frontal and temporal lobes, which in living humans are associated with such higher cognitive processes as taking initiative and advance planning.

If a hominid with a brain no larger than a chimp's could make stone tools, organize hunting parties to kill and butcher large and dangerous prey, and build fires, one has to wonder why our ancestors didn't start engaging in these practices sooner than they apparently did. It is not until 2.6 million years ago—nearly 4 million years after the first hominids emerged—that our predecessors began making stone tools (that's as far back as they go in the fossil record, anyway). Lucy and her cohorts do not appear to have had a material culture. The earliest stone tools for which we have associated hominid remains are the 2.35-million-year-old Oldowan tools my team found along with an early *Homo* maxilla at the Maka'amitalu locality at Hadar. And control over fire came much later still: The earliest known evidence for this dates to 790,000 years ago. It must be the wiring of the brain—not the sheer volume of it—that determines how innovative we are.

Falk's work has not silenced the critics, however. In a paper published in the May 19, 2006, *Science,* a team led by Robert Martin of Chicago's Field Museum argued that LB1's brain is far smaller than expected for its body size if it descended from *H. erectus,* based on what is known about dwarfing in other mammals. Referencing scaling data available for other mammals, Martin's team argues that a descendant of *H. erectus* with a brain the size of LB1's should have a body mass of merely 11 kilograms—the size of a small monkey, not a chimpanzee or an australopithecine. The authors also asserted that Falk used the wrong microcephalic in her comparative sample—a boy who died at the age of ten and with a brain volume of only 260 cubic centimeters is the smallest microcephalic on record. Microcephaly is a symptom that occurs in over two hundred disorders and looks very different in the details depending on the disorder it accompanies. Considering that LB1 was an adult when she died, Falk should have chosen a microcephalic with a milder condition that enabled it to survive into adulthood, Martin admonished. His team looked at the endocast of a thirty-two-year-old woman from Lesotho and an adult male from India. Both bespoke brain sizes comparable to LB1's, he found, and small size aside, both looked surprisingly normal. For her part, Falk defended her team's use of the ten-year-old's endocast, noting that in severe cases of microcephaly, individuals typically die within the first few years of life. She has since examined more microcephalics and none, she reports, look like LB1. Paleoneurologist Ralph Holloway of Columbia University has also studied LB1's endocast. He found several features in addition to its small size that could indicate pathology, including a pronounced flattening known as platycephaly. But he was unable to spot these features in any of the twenty microcephalic endocasts he examined.

Meanwhile, further clues to LB1's ancestry and appearance have come from the postcranial skeleton. William Jungers has scrutinized the limb bones of LB1 and found them to be as thick and sturdy as those of chimps and Lucy. No mere midget, she must have been very strong. And judging from her pelvis, she certainly walked bipedally. She probably did not walk or run like us, however. Jungers has determined that LB1 had incredibly short legs—so short, in fact, that her feet are nearly as long as her shinbone. At a presentation given to the American Association of Physical Anthropologists in April 2008, Jungers explained that LB1 lacks the features that made *H. erectus/H. ergaster* and their descendants good runners, such as long legs, short toes, and an arched foot. "I'm pretty sure they could dash to safety from menacing Komodo dragons (their only remotely possible predator), even run down a baby *Stegodon* for lunch," he remarks. "But they couldn't run a marathon or win the gold medal in the 220-meter dash."

Neither does LB1's upper-limb anatomy resemble ours. Studies of the shoulder anatomy, conducted by Susan Larson, have revealed similarities to *H. erectus*, not *H. sapiens*. The humerus (upper arm bone), for example, is straight, not twisted like our own. And her collarbones are rather short. Larson's research indicates that this is a primitive characteristic, retained from a time when hominids were still transitioning from a shoulder built for walking on all fours to one possibly optimized for throwing and/or upright endurance running. LB1 would not have excelled at either of these latter activities.

One of the most convincing pieces of evidence for the hobbit being a new hominid species has come from Matthew Tocheri, now a researcher at the National Museum of Natural History in Washington, D.C. During his tenure as a student in our Ph.D. program at Arizona State University, Tocheri studied the hand and wrist bones of hominids through time for his dissertation. As part of his research, he examined casts of LB1's wrist bones. The specimen preserves three key bones in the left wrist—the trapezoid, scaphoid, and capitate—all in good condition. In early and contemporary modern humans and our close relatives, the Neandertals, the trapezoid is shaped like a boot, and the other wrist bones articulate with it accordingly. In contrast, nonhuman primates and australopithecines like Lucy have wedge-shaped trapezoids and a correspondingly different arrangement of the scaphoid and capitate. When Tocheri compared LB1's wrist bones with those of other hominids and nonhuman primates, he found that she had a wedge-shaped trapezoid. "It blew my mind," he recollects. "I didn't think the wrist would say anything about whether [LB1] was a distinct species or not." But overall, her wrist looks most like the wrists of African apes and fossil humans dated to more than 1.7 million years, including Lucy. Importantly, Tocheri—who published his findings in the September 21, 2007, issue of *Science*—included a pituitary dwarf and a pituitary giant in his comparative sample of modern humans. Because these individuals exhibit normal-looking modern wrist bones, despite having severe growth disorders, he was able to argue persuasively that LB1's primitive wrist morphology resulted from normal growth and development, not some sort of disease. He further notes that even modern wrists that do bear signs of pathology don't look like LB1's. "A lot of things can go wrong in wrist development," Tocheri comments. "But there are no examples of [a modern human] developing a normal-looking African ape wrist."

Tocheri's findings offer strong support to the hypothesis that LB1 evolved from a human ancestor that departed Africa before the modified wrist of Neandertals and anatomically modern humans evolved. The earliest evidence of the more advanced wrist comes from a specimen of *H. heidelbergensis* from Spain that dates to 800,000 years ago. Tocheri postulates that the modern wrist took shape sometime between 1.8 million years ago and 800,000 years ago. This new arrangement optimized the distribution of forces across the wrist and hand, which may have facilitated tool manufacture and use. If so, that could help explain why the tools of early modern humans and Neandertals are so much more refined than those of their predecessors.

Although Tocheri's work undermines the idea that LB1 is a diseased modern human, the identity of the hobbits' ancestors remains uncertain. Brown and Morwood initially suggested only that the little Floresians were the descendants of *H. erectus*, which inhabited Southeast Asia between 1.8 million years ago and 25,000 years ago. If the hobbits were descended from local *H. erectus*, that would make good biogeographic sense. But LB1 and her cohorts don't look particularly like East Asian *H. erectus*. They do, however, somewhat resemble *H. erectus* individuals from Dmanisi, who have a similarly primitive shoulder morphology and generally smaller body and brain proportions, though none are so petite as the hobbits. On the other hand, LB1's small size and archaic-looking shoulder, wrist, and pelvis also call to mind Lucy. Those similarities raise the possibility of two additional ancestry scenarios. Maybe LB1's ancestors came from the same stock as the Dmanisi folks, and when they arrived on Flores—or another small island in the Indonesian archipelago—they evolved even smaller proportions. Or perhaps the hobbit forebears were more primitive hominids—*H. habilis* or even *Australopithecus*—who set forth from the African continent and penetrated Southeast Asia.

Each of these models has its strengths and weaknesses. If the hobbits descended from the Dmanisi folks, the biogeography works, but one still has to explain how brain size further diminished. If, on the other hand, the hobbits descended from *H. habilis* or an australopithecine, their brain and body proportions are accounted for, but an established view of hominid biogeography is upended: Neither *H. habilis* nor australopithecines have ever been found outside of Africa, and it is widely assumed that they never made it off the continent. *Homo erectus* is believed to have been the first hominid to leave the motherland. Only the discovery of hominid remains linking the hobbits to hominids from more than 2 million years ago can resolve this question. But now researchers have a reason to look for remains. It's a tall order, but one that is breathing new life into the field. "The evidence points to a really bright and exciting future for paleoanthropology," Tocheri enthuses. "It tells us we should find a whole lot more, not only on Flores but on the surrounding islands and on the mainland."

What of the claim that LB1 is simply a diseased modern human? Recently, researchers have proposed a number of disorders to explain LB1's odd features. One team suggested Laron syndrome, a genetic disease that causes insensitivity to growth hormone. A second group proposed another genetic condition known as microcephalic osteodysplastic primordial dwarfism type II (MOPD II), which leaves individuals with small bodies and small brains, but near-normal intelligence. And in March 2007, a third team posited that LB1 suffered from myxoedematous endemic cretinism, a disorder that stems from prenatal nutritional deficiencies that disable the thyroid.

But the assault wasn't over. Days after the publication of the cretinism paper, hobbit skeptics seemed to get another boost, this time from the discovery of tiny modern human bones ranging in age from 1,400 to 2,900 years old in two caves in Palau, Micronesia. Lee Berger of the University of the Witwatersrand in Johannesburg, who found the remains, and his colleagues reported that the Palauan bones, in addition to being small, exhibit traits typically associated

with earlier *Homo,* such as pronounced browridges and a nonprojecting chin. The hobbits, too, have these characteristics, which have
been used to help make the case that they belong to a new species.
But Berger's team argues that these features may just be correlates
of small size. If so, that would support the theory that LB1 is merely
a small modern human with a disease that left it with a teeny brain,
among other peculiar traits.

The hobbit team has denounced all of these arguments. "The
evidence offered by the critics in support of microcephalic dwarfs,
Laron syndrome, cretins, etc., is absolutely dreadful," Brown scoffs.
All the critics have to do is produce a skeleton that replicates the
combination of traits in LB1, he says. But that hasn't happened. As
for the Palau findings, it turns out the alleged browridges were
merely calcite deposits from the caves in which the bones were
found. And some of the individuals Berger described appear to be
juveniles whose features were still developing, which might explain
the nonprojecting chins.

Most damning of all is a report published in August 2008 by
anthropologist Scott Fitzpatrick of North Carolina State University
and his colleagues. Their analyses of complete bones and skeletons
from sites dating back nearly 3,500 years has shown that early
Palauans were of normal stature. According to the authors, Berger
incorrectly extrapolated body weight from the fragments he found.

"Other human groups scattered around the world are also just
as small or smaller than these fragments from Palau," Jungers says.
"But none of them anywhere in the world are as short as the various
individuals of *Homo floresiensis*." Moreover, none of them possess
LB1's primitive features, including her short legs, her apelike wrist
bones, and her flaring pelvis, which bears a striking resemblance to
an australopithecene's. In fact, when Jungers articulated LB1's hip
bone with a cast of Lucy's sacrum, the fit was perfect, and the hobbit's hip looked remarkably like Lucy's.

Many observers had hoped that DNA might settle the issue once and for all. But so far, sequences of mitochondrial DNA obtained from LB1 look totally modern. The best explanation for such a result, says Svante Pääbo of the Max Planck Institute for Evolutionary Anthropology, who conducted the analysis, is that the sample was contaminated with the DNA of someone who handled the remains.

Still, most experts have embraced the notion that LB1 and the other individuals that have been recovered so far at Liang Bua do in fact constitute a new branch of the human family tree. There is a long and storied tradition in paleoanthropology of skeptics dismissing unexpected hominid finds. The discovery of the first Neandertal skeleton in 1856 is one example—doubters argued that it was a crippled Cossack horseman. And in an even more striking example of history repeating itself, when Eugène Dubois introduced Java Man to the world as a new hominid species in 1891, British paleontologist Richard Lydekker rejected the claim, saying it was instead—wait for it—"a microcephalic idiot."

There are, of course, actual examples of diseased individuals in the fossil record. One Neandertal skeleton is now recognized to have the hallmarks of arthritis (though no modern scholar disputes that it is in fact a Neandertal). And the celebrated *H. erectus* skeleton known as the Turkana Boy shows signs of having had a developmental pathology known as neural stenosis, as described in the preceding chapter. Most hominids of prehistory probably lived hard and died young compared to the cushy lives we in the developed world lead. But by and large, when assessing hominid remains, it's fair to assume until proven otherwise that any given specimen is not an outlier but representative of the population from which it came. At Liang Bua, Morwood's team has recovered the remains of a dozen individuals, which nearly qualifies as a population in paleoanthropology. Although none are nearly so complete as LB1 (some are represented by isolated bones), all appear to have been small. (According to Jungers, LB1 was the largest of the bunch.) The problem is, LB1 is the only individual whose skull has been recovered.

Thus the possibility remains that the other hobbits had normal brain sizes. This, in fact, is exactly what the skeptics predict. But if future excavations in the cave—which has since been reopened—turn up even one more small skull, the microcephaly theorists will be proved wrong. In the meantime, the burden of proof is on the critics to find a pathological condition that replicates the unique morphology of LB1—no simple feat, considering that microcephaly is a symptom that occurs in hundreds of disorders.

Taken at face value, the picture emerging from Liang Bua is every bit as colorful as the scenes out of Tolkien's books: miniature hominids surviving on a remote tropical island by hunting pygmy *Stegodon* and giant rats, and dodging dragons. Eventually, however, the little hobbits disappeared, possibly the victims of a mighty eruption of one of the island's many volcanoes. They may live on in spirit, though. Modern-day residents of Flores tell stories about a small, hairy creature they call *ebu gogo*, "the grandmother who eats everything." According to legend, the *ebu gogo* had a lopsided gait, murmuring speech, and—as the name suggests—an insatiable appetite. Oral histories hold that the *ebu gogo* was still in existence when the Dutch colonized Flores in the 1800s. And when Carolus Linnaeus—the so-called father of taxonomy—coined the genus *Homo* in his tenth edition of the *Systema Naturae* (his classification of living things) in 1758, he named *H. sapiens* and a second species, *H. sylvestris*. He described the latter as a nocturnal, cave-dwelling human from the island of Java in Indonesia. Today *H. sylvestris* is believed to be mythical, but the discovery of the hobbits on nearby Flores begs the question of whether Linnaeus's second species might have been more than a fantasy. Could the *ebu gogo, H. sylvestris,* and the hobbits be one and the same? Food for thought indeed.

The Neandertals

Of all the species that Lucy and her kinsmen ultimately gave rise to, none has engendered such fervent debate as our closest cousins, the Neandertals. It has been this way ever since the very first skeleton of this extinct human emerged from a limestone cave known as Feldhofer near Düsseldorf, Germany, in 1856. The quarry workers who happened upon the remains believed they had found a cave bear and collected the heavily built skull and various other bones to show to local schoolteacher and amateur naturalist Johann Carl Fuhlrott. Fuhlrott recognized the specimen as an ancient human, and together with anatomist Hermann Schaaffhausen published a paper announcing the discovery the following year. But it was the Irish geologist William King who actually assigned the specimen to the species *Homo neanderthalensis,* the first fossil hominid ever named. (Around 1900 German orthography changed and the silent *h* in certain words was dropped, hence the usage here of Neandertal instead of Neanderthal. The designation *H. neanderthalensis* remains the same, however.) King's was an audacious assertion, for evolution was hardly a widespread idea at the time. Predictably, naysayers quickly emerged. Most notable among them was the respected medical doctor and anthropologist Rudolf Virchow, considered the father of pathology, who countered that the remains were simply those of a modern human whose odd features could be attributed to childhood rickets, followed by arthritis in adulthood. And thus the infamously contentious field of paleoanthropology was born.

Subsequent Neandertal discoveries from France, Belgium, and other parts of western Europe over the next few decades confirmed that King and others were correct in recognizing the quarrymen's find as an ancient hominid distinct from modern Europeans. They also fueled another of King's notions—namely, that this hominid was a subhuman barbarian. French paleontologist Marcellin Boule reached a similar conclusion to King's, based on a fossil discovered at La Chapelle-aux-Saints, in southwestern France. From his analyses of the specimen, conducted between 1911 and 1913, Boule determined that the individual had divergent big toes, as chimps do, and walked stooped over and bent-kneed, with his head sticking forward. The Neandertals, in his view, were dim-witted, crude creatures who slouched and shuffled their way to extinction—more advanced than an ape, but vastly inferior to modern humans in every way. It was a most unflattering portrait.

Half a century later, reanalysis of the La Chapelle Neandertal revealed that it was the skeleton of an old male who had suffered from severe arthritis—hardly representative of the average Neandertal. It became apparent that, contrary to Boule's assessment, Neandertals carried themselves upright as we do. Nonetheless, the perception of these hominids as primitive protohumans stuck.

Today the Neandertals are arguably the best-studied member of the human family after *Homo sapiens,* thanks to the relatively rich record of their remains. We now know that they ruled Europe and western Asia for more than 200,000 years, persevering through the harshest Ice Age conditions endured by any human ancestor before disappearing around 30,000 years ago. Yet our closest kin remain in many ways mysterious. Just how much like us were they? What brought about their demise? Do any of their genes live on in us? These questions have captivated scholars for decades. And though we still cannot answer them with certainty, recent findings have shed considerable light on the lives and times of these bygone Eurasians.

The story of the Neandertals begins with a hominid called *Homo heidelbergensis*. Anatomist Otto Schoentensack coined the name in 1908 for a massive mandible found just outside the German village of Mauer, near Heidelberg. But in the years that followed scholars came to view the specimen, with its mix of primitive and modern traits, as an "archaic *Homo sapiens*." It wasn't until after the discoveries decades later of similar fossils in places as far-flung as Zambia, Ethiopia, Israel, Morocco, and many sites in between that experts began to embrace the designation *heidelbergensis*. Today most researchers consider *heidelbergensis* to be a distinct species, descended from *H. ergaster*, that lived from roughly 700,000 to 200,000 years ago, during the Middle Pleistocene. In Africa, it gave rise to anatomically modern humans. But in Europe and western Asia it spawned another group of large-brained hominids, the Neandertals. (There are, of course, other interpretations of this section of the fossil record—this is paleoanthropology, after all. Some investigators have proposed that the bones from a site called Gran Dolina in Sierra de Atapuerca, Spain, constitute another Middle Pleistocene hominid species, *H. antecessor*, that they believe was the common ancestor of modern humans and Neandertals.)

Closely related though we are to Neandertals, their form is recognizably different from our own—even to the untrained eye. The skull looks somewhat like a football in being long and low, with the rear of the cranium bearing a distinctive bulge or "occipital bun," as it is termed. And the face exhibits, among other traits, a thick, double-arched browridge, wide nasal opening, and a chinless mandible. For its part, *H. sapiens* possesses a globular skull, a smooth brow, a narrower nose, and a jutting chin. The Neandertal body is likewise distinctive. In addition to being short and squat, it has a bell-shaped rib cage and broad hips, which effectively meant that this hominid didn't have a waist. We modern humans, in contrast, have a taller, slimmer frame, with a tapered rib cage and narrower pelvis.

This much about Neandertal biology has been known for decades. Researchers concur that some of the traits evident in Neandertals—the recessed chin, for instance—are probably just holdovers from *heidelbergensis*. Others may have arisen randomly, and because they were harmless became fixed in the population through genetic drift. But a number of their features were likely adaptations to the cold climate in which Neandertals lived. The large nose, for example, may have facilitated the moistening and warming of that frigid incoming air. The stocky, muscular Neandertal build, meanwhile, helped conserve heat. These denizens of the Ice Age needed all the help they could get: During glacial times, windchill temperatures commonly dipped below 0 degrees Fahrenheit. Interestingly, the Neandertals had very thick cortical bone, the building of which requires substantial amounts of calcium. Among Inuit and Aleut populations of the Arctic, the bone is very thin and porous, leaving these people prone to osteoporosis, among other health issues. The problem is they don't seem to be able to get enough sunlight for vitamin D synthesis, which is what facilitates the intestinal absorption of the calcium needed to build strong bones. Neandertals, too, lived in high latitudes with less intense solar radiation and shorter days in the long winters. How did they obtain the calcium needed to build such robust bones? Considering that modern-day Europeans have evolved light skin, which absorbs more of the ultraviolet B radiation that supports vitamin D synthesis, it is likely that the Neandertals, too, evolved light skin. (The Inuits and Aleuts arrived in the Arctic only recently, and therefore haven't had sufficient time to evolve lighter skin.)

Specialists have long been divided over whether the distinctive Neandertal features warrant placing these hominids in a species separate from *H. sapiens*. Proponents of the idea that Neandertals were

members of our own species argue that many of the traits that characterize Neandertals occur in the early modern Eurasians that followed them, indicating that the indigenous Neandertals interbred
with the incoming modern humans, which would make the two
groups members of the same species. Naysayers, however, maintain
that these similarities, such as prominent brows, are rare and can be
explained as having been inherited from a common ancestor that
possessed those features.

Contributing to the problem is the fact that sometimes experts
disagree over not only interpretation but observation. Consider the
case of a fossil known as Lagar Velho 1. Discovered in a rock shelter
in central Portugal's Lapedo Valley, this mostly complete skeleton is
that of a child who died around 24,500 years ago and was buried in
the Gravettian style known from modern human sites of comparable
antiquity across Europe. Neandertal authority Erik Trinkaus of
Washington University and his colleagues, who formally described
the remains in 1999, argue that Lagar Velho 1 is a modern child, as
evidenced by its prominent chin, short face, and minimal brow
development, among other characteristics. But they claim that it
also exhibits a number of Neandertal traits, such as strongly developed pectoral muscles and lower leg bones that are short and stout.
Considering that the Neandertals are believed to have disappeared
thousands of years before this youngster's time, it couldn't have
been the "love child" of a modern human and a Neandertal. Rather,
Trinkaus and his team reason, Lagar Velho 1's mosaic of modern
and archaic traits must have been the product of considerable mixing between the indigenous Iberian Neandertals and the modern
humans who arrived sometime after 30,000 years ago. Other scientists interpret the skeleton quite differently. Ian Tattersall of the
American Museum of Natural History and Jeffrey Schwartz of the
University of Pittsburgh, for example, have countered that it was
probably just a "chunky Gravettian child." They see no need to
invoke the Neandertals to explain the child's morphology. I concur

with Tattersall and Schwartz's view and see the Lagar Velho child as a cold-adapted *H. sapiens*. After all, today's Eskimos are short and squat with short, stout shins, undoubtedly an adaptation to conserving body heat in an Arctic environment.

Similar disputes have erupted over many other fossils. But in the last decade, researchers have begun tapping a new source of data that could help us settle this debate once and for all: ancient DNA. The work is ongoing, but scientists have been able to retrieve and sequence ancient DNA from both Neandertal and early modern human fossils. Most of the DNA analyses to date have focused on mitochondrial DNA (mtDNA), and those studies concluded that Neandertals did not mingle with moderns. But mtDNA constitutes only a very small part of a person's genetic makeup, and analyses of ancient mtDNA have left many researchers wondering what the rest of the Neandertal genome looks like. To that end, in what is shaping up to be a stunning technical feat, researchers led by Svante Pääbo, an expert in ancient DNA, of the Max Planck Institute for Evolutionary Anthropology are reconstructing a rough draft of the entire Neandertal genome. Already the work has yielded some provocative findings. Neandertals apparently had the same variant of *FOXP2*—a gene linked to speech and language—that modern humans have, hinting that they may have communicated as we do. And analysis of part of a gene known as *MC1R* that determines skin and hair color suggests that Neandertals had the same distribution of hair and skin color as modern Europeans do, with around 1 percent of them being redheads with pale skin and freckles. So far, however, the team has not found any evidence of mixing between Neandertals and moderns.

The issue of whether or not Neandertals belong on their own branch on the human family tree is an important one. But it's by no means the only question researchers have been trying to answer. Much work has gone into elucidating Neandertal behavior, which was long misunderstood. "Neandertals used to be considered incomplete versions of ourselves," observes archaeologist John Shea of Stony Brook University. And while this temporarily made it easier to understand why they eventually disappeared, investigators soon began finding signs that Neandertals were far cleverer than they were being given credit for.

For one thing, they figured out how to stay warm during glacial periods. Their cold-adapted biology alone could not enable them to survive the deep freeze. In fact, paleoanthropologist Leslie Aiello, now at the Wenner-Gren Foundation, and physiologist Peter Wheeler of Liverpool John Moores University have calculated that Neandertal bodies were only slightly better equipped than those of modern humans to withstand cold. Under the most frigid conditions, they determined, Neandertals would have had to wear clothing in order to survive. Fur, no doubt, was always in vogue. But they would have to have tailored the hides to prevent the loss of body heat. Although no needles have ever turned up at Neandertal sites, they may well have used bone awls to fashion their garments. As for shelter, caves would have afforded protection from the elements. But some Neandertals appear to have built their own dwellings: At a site known as the Grotte du Renne in Arcy-sur-Cure, France, archaeologists found remnants of post holes—a sign that residents had built a structure from mammoth tusks that they probably draped with tanned hides to form a cozy tent of sorts. A blazing campfire, too, would have helped keep the lethal cold at bay.

The colder temperatures that Neandertals had to contend with presented more challenges than just staying sufficiently warm, however. Finding food in this environment was difficult, too. Unlike their counterparts in toastier climes, which have a cornucopia of plant foods on offer throughout the year, Neandertals would have found themselves with few such comestibles available to them, particularly from autumn through spring. Most edible plants just couldn't withstand the frigid temperatures. So the Neandertals did the only thing they could do: They adopted a diet based largely on meat, specifically big game.

That the Neandertals subsisted primarily on animal flesh is by now well established. But archaeologists continue to debate exactly how Neandertals obtained meat. Some have argued that they were inept hunters, capable only of bringing down small and/or young antelope and the like. As such, they would have been obligated to thieve kills from the local carnivores. Others have proposed that they were opportunistic scavengers, stealing when it was convenient, and hunting—albeit less effectively than modern people—the rest of the time. And still other experts consider Neandertals to have been as accomplished as any modern hunter. Which scenario is the correct one? Insights have come from studies of contemporary settings. Research conducted in the 1980s by Robert Blumenschine of Rutgers University found that even in the rich tropical grasslands of northern Tanzania, scavengeable food is scarce most of the year. Although we lack comparable data for colder environments, we know that they have fewer large mammals than tropical ones do. Given how utterly dependent Neandertals apparently were on meat to survive, and how unreliable scavenging is as a strategy for meat acquisition, it seems to me that they must have been skilled hunters to sustain themselves over the barren winter months.

Theirs was not an easy way to make a living. Neandertal skeletons often bear signs of trauma. In fact, Neandertal expert Erik Trinkaus found that these hominids had trauma patterns comparable to those of rodeo athletes, with high numbers of injuries to the head and neck. This similarity, he concluded, suggested that the Neandertals had frequent close encounters with prey. So far as we know from their stone tools, these hominids did not make the small points that one could turn into projectile weapons. Instead, they probably used handheld spears to bring down their quarry. The muscular Neandertal physique no doubt facilitated this close-range combat. But because muscle tissue is energetically expensive to maintain, it also meant that these hominids had to eat more than their punier kin. Based on their studies of Siberian reindeer-herding populations known as the Evenki and the Inuit populations of the

Canadian Arctic, biological anthropologists William R. Leonard and Mark Sorensen of Northwestern University estimate that Neandertals had to consume as many as 4,000 kilocalories a day to survive, compared to the 2,600 kilocalories a 160-pound American male with a typical urban lifestyle needs.

So focused were the Neandertals on killing these large animals that they may have had rather unorthodox hunting parties. Modern hunter-gatherer societies have a division of labor wherein males undertake the hunting of large animals while women collect small game and sew garments, among other tasks. In a paper published in *Current Anthropology* in 2006, University of Arizona archaeologists Mary Stiner and Steven Kuhn observed that sites occupied by modern humans between 45,000 and 10,000 years ago include remains of small animals and bone needles, which they take to indicate that the men and women there engaged in different activities. Neandertal sites, on the other hand, contain no such remnants. Stiner and Kuhn theorize that because Neandertals pursued a narrower dietary strategy than do modern foragers, women's work may have been one and the same with men's work, which is to say both sexes engaged in the pursuit of large game. Healed bone fractures in the skeletons of female Neandertals support this suggestion. Children, too, may have joined the hunt, the researchers posit, working alongside the women as beaters.

Fearsome though the Neandertals must have been to routinely tackle large prey with only handheld wooden spears, they also had a sensitive side. In the late 1950s excavators working at the cave site of Shanidar in northeastern Iraq found the 60,000-year-old skeleton of an elderly male, Shanidar 1, who had endured a number of serious injuries during his life. He had a withered upper arm bone, a crippled right leg, and a fractured right foot. And he had suffered a blow that crushed his left eye socket, which probably left him blind in that eye. These injuries surely left him unable to fend for himself. Yet all of them show signs of healing, meaning that he survived for some time after sustaining the damage. He must have relied on other clan members to bring him food, keep him warm, and protect him from predators. A similar case of Neandertal compassion surfaced in 2001, when archaeologists reported on their discovery of a nearly toothless mandible at a rock shelter in southeastern France called the Bau de l'Aubesier. Dated to between 169,000 and 191,000 years old, the bone belonged to an adult who had somehow lost his teeth, which in turn exposed his gums to infection that eventually led to bone disease in his jaw. Although he would have been largely unable to chew food, he still managed to survive for at least six months in this state. The discovery team posits that his comrades provisioned him with naturally soft foods, such as fruits, or tougher foods that they processed for him so that he could eat without chewing.

Not all Neandertal foraging tactics were the same, however. Mary Stiner and Steven Kuhn have found that at sites in west-central Italy, prior to 55,000 years ago, Neandertals ate a lot of tortoises, shellfish, and scavenged game. After 55,000 years ago, however, they shifted to a strategy that was heavy on hunting ungulates, such as red deer, fallow deer, and ibex. These findings put paid to another common misperception about Neandertals; namely, that they were stuck in their ways, unable to adapt to change. Stiner and Kuhn aren't the only ones to find evidence of Neandertal adaptability. University of Pennsylvania archaeologist Harold Dibble observes that Neandertals tailored their toolmaking practices to the resources available to them, conserving materials when they were in short supply and making the best of whatever they had at their disposal. And Shea notes that when Neandertals moved into the Levant, they began making tools they had never made before, such as sophisticated stone-tipped spears. The notion that Neandertals were unable to adapt "is not even a little bit true," Dibble asserts.

But it may be that modern humans were better than Neandertals at adjusting to changing circumstances. According to one popular theory, a greater capacity for cognition, possibly including language, gave moderns an evolutionary edge over Neandertals. This enhanced capability seemed to kick in sometime around 40,000 years ago, at which point early modern Europeans began making art, decorating their bodies, and fabricating complex bone and antler tools—a cultural tradition broadly referred to as Upper Paleolithic. Neandertals, however, continued to make mostly the same Middle Paleolithic artifacts they had been making all along. Yet there were hints here and there that Neandertals had similar stirrings of symbolic thought.

Researchers have argued that a number of Neandertals were deliberately buried by their contemporaries, including skeletons from the sites of La Chapelle-aux-Saints, La Ferrassie, and Roc de Marsal in France; Amud and Kebara in Israel; and Shanidar. And some of these interments, it has been argued, were ceremonious. At Shanidar, soil samples collected in the 1960s from the grave of another elderly male, known as Shanidar 4, were found to contain pollen from a variety of colorful flowers, many with medicinal properties. The discovery of the pollen conjured images of Neandertals strewing flowers on the graves of their dearly departed and recast these hominids as the original flower people. It was, after all, the height of the hippie movement. But subsequent investigations raised the question of whether some other agent might have deposited the pollen—wind, for example, or a gerbil-like rodent known as the Persian jird that stores seeds and flowers in its burrows. Other putative examples of Neandertal burial rituals are equally nebulous. At La Chapelle-aux-Saints, leg and foot bones belonging to a bovid were recovered near a Neandertal skeleton, prompting some experts to posit that the mourners had made an offering of food for the deceased. Graves at La Ferrassie, meanwhile, contain implements that have been interpreted as grave goods, possibly indicating that these hominids believed in an afterlife. At a site called Teshik-Tash in Uzbekistan, a child's grave was ringed with mountain goat horns driven vertically into the ground. And at several sites, the skeletons were found in a flexed position, which some observers have taken to mean that the mourners positioned the deceased to mimic the fetus's posture in the womb.

Skeptics, however, assert that in all of these cases, the supposedly symbolic items were commonplace in Neandertal residences and probably just happened to be swept into the graves. Some experts are doubtful that Neandertals buried their dead at all, never mind the ritual element. According to Dibble, who has been working at the Roc de Marsal site, recent geologic analyses have revealed natural features in the cave that could mimic burial. And other sites containing alleged Neandertal graves—all of which were excavated long ago—lack the necessary documentation to demonstrate convincingly that the bodies were interred intentionally, he cautions. As for the fetal position of some Neandertal specimens, one must ask how Stone Age people would have known what the fetus looked like in the womb. A more parsimonious explanation is that making the corpse compact meant that they could dig a smaller hole.

Burials of early modern Eurasians, in contrast, are unequivocal and leave no doubt as to the emblematic nature of the grave goods. A particularly magnificent example comes from the Russian site of Sunghir. There, a sixty-year-old man was buried wearing twenty-five bracelets, a pendant, and a tunic decorated with thousands of beads made of mammoth ivory some 24,000 years ago. Another grave at the same site contained two juveniles buried head to head with even more beads and other grave offerings. According to archaeologist Randall White of New York University, an expert on ancient personal ornaments, each of the more than 13,000 beads contained in these burials would have taken an hour to make. That these labor-intensive items were interred along with the dead—and in such vast quantities—makes clear that their association was intentional.

We may never know whether Neandertal burials, if they occurred, were ritualized or whether they merely constituted a pragmatic way of disposing of corpses that might otherwise attract unwelcome scavengers. But there are tantalizing clues to suggest that at least some Neandertals engaged in other kinds of symbolic behavior. Dozens of sites have yielded chunks of pigment, hinting that Neandertals decorated their bodies. A particularly compelling example of this behavior, announced in 2008, comes from Pech de l'Azé in France, where Francesco d'Errico of the University of Bordeaux and his colleagues recovered hundreds of pieces of black manganese with wear patterns indicating that they had been used to draw straight black lines. Because body painting is a means of communicating with symbols, experts commonly take it to be a proxy for the ultimate symbolic behavior, language.

Furthermore, at a handful of sites in France and along the Pyrenean and Cantabrian mountain ranges, researchers have found an advanced Neandertal cultural tradition called the Châtelperronian that includes complex stone and bone implements, decorated objects, and body ornaments—all items once believed to be the exclusive purview of modern humans. France's Grotte du Renne, for one, has yielded a number of pierced animal teeth that would have been suspended from a cord and worn as pendants, as well as bone awls and stone blades.

Yet the Châtelperronian evidence, too, has drawn doubts. Early modern humans had a comparable industry known as the Aurignacian that appears at a number of the same sites that contain Châtelperronian levels. Some researchers have thus argued that disruption of the archaeological layers at these sites mixed in artifacts made by moderns with the Neandertal relics. There are also those who contend that the Neandertals acquired these objects through trade with moderns, in a sort of Stone Age swap meet. It has even been proposed that Neandertals stole the items from their modern neighbors or collected their discarded goods. Other skeptics grant that the Neandertals manufactured the advanced-looking items, but contend that they did not invent them on their own. Noting that Neandertals only began crafting Châtelperronian tools and ornaments after anatomically modern humans arrived in western Europe, archaeologist Paul Mellars of Cambridge University contends that in fabricating sophisticated objects, these late Neandertals were simply aping the ways of the newcomers. His interpretation has not gone unchallenged. João Zilhão of the University of Bristol and Francesco d'Errico have long maintained that the Neandertals were already evolving toward behavioral modernity before they ever came into contact with moderns. At sites that contain both Châtelperronian and Aurignacian remains, the former always underlie the latter, they insist. Furthermore, the Châtelperronian objects reflect a different method of manufacture. In the case of pendants, which were made by modifying bear and wolf teeth, among others, the Neandertals preferred to etch a furrow around the tooth root so that a string could be tied around it for suspension, although sometimes they would puncture the tooth. The Aurignacians, in contrast, favored scraping the tooth thin and then piercing it. And as for the cutting-edge Châtelperronian tools, they, too, reflect manufacturing techniques quite distinct from those employed by the Aurignacians, and seem to derive from early toolmaking practices.

Neandertals lacked the heightened cognitive abilities of moderns, but they were certainly not dullards. So why, after dominating Eurasia for millennia, did these kings of the Ice Age vanish? The latest Neandertal sites we know of occur in the Iberian peninsula and date to close to 30,000 years ago. Two things happened leading up to this point: Anatomically modern humans from Africa spread throughout Eurasia, and climatic instability began rendering the environment ever more hostile to human habitation. It is surely not by chance alone that the decline of the Neandertals coincided with these events, particularly the arrival of moderns in the region. According to Chris Stringer of the Natural History Museum in London, moderns were probably a little more innovative and capable of dealing with rapid environmental shifts, and they probably had bigger social networks. Others, including Fred H. Smith of Northern Illinois University, contend that the Neandertals were absorbed into the larger modern population. Moving into the realm of sheer speculation, it has even been proposed that moderns killed off the Neandertals or exposed them to a deadly infectious disease.

My colleague Curtis Marean, an archaeologist at the Institute of Human Origins, has what I consider to be the most compelling hypothesis for how early moderns rose to dominance in Eurasia. Traditionally, investigators have viewed Neandertals from a Eurasian perspective, because that's where their remains are found. Curtis approaches the question somewhat differently from researchers who focus on the Eurasian record because he works at early modern human sites in Africa. He has thought a lot about the factors that molded those moderns who moved into Neandertal territory, and in so doing has identified important differences in the behavioral ecology of the two groups. As Curtis sees it, when Neandertals and modern humans split from their last common ancestor around 400,000 years ago, Neandertals ended up on a trajectory of evolution in the colder setting of Eurasia that was quite different from the trajectory that modern humans found themselves on in tropical Africa. Over time, Neandertals adopted a very specialized foraging system that was focused on larger animals on the landscape. The associated costs and risks of this tactic were high, but so, too, were the returns. Specialized predators do very well in certain circumstances, and this strategy stood the Neandertals in good stead for millennia. But specialists get into trouble when their main food item starts to decline or when another predator moves in and starts to compete with them for the specialized animal. Indeed, a well-established rule of ecology holds that generalists always win out over specialists.

Enter *H. sapiens,* who evolved under a much more complex set of conditions in their tropical birthplace, where a variety of edible plants and smaller game abounded. Here natural selection favored the development of a hunter-gatherer society with an array of strategies to exploit the diversity of available foodstuffs. Some of these strategies were quite sophisticated: exploitation of plant foods, especially, demanded in-depth knowledge of the landscape, including the seasonality of particular plants. The ability to mentally map food availability no doubt fostered the development of enhanced communication skills. Thus when *H. sapiens* arrived in Europe, it was well equipped both behaviorally and cognitively to outcompete Neandertals. Yet it is nonetheless remarkable that members of a tropical species migrated into a totally foreign environment and, despite being unfamiliar with the geography and having no knowledge of local animals or plants, managed to beat out the Neandertals on their own turf after they had lived there for hundreds of thousands of years.

"Modern humans come into western Eurasia eating anything that moves," Curtis says. "Neandertals are going to go extinct because of the competition." It's not hard to see why. Large animals tend to exist in relatively small numbers in any given place—the landscape just can't support that many of them. Thus early moderns wouldn't have had to take many of these beasts to significantly impact Neandertal hunting success.

Studies of Neandertal bone chemistry support Curtis's model. After analyzing the chemical composition of samples of Neandertal bone and the bones of other mammals, Hervé Bocherens of the University of Tübingen and his colleagues concluded that European Neandertals often targeted some of the largest and most dangerous animals around, including woolly mammoths and woolly rhinoceroses. It was a risky tactic, to be sure. No human group alive today habitually hunts elephants or rhinos for their meat—it's just too dangerous. Why, then, did Neandertals preferentially pursue these animals instead of, for example, the horses and reindeer that lived in their midst? The answer, it seems, is that hyenas and other predators already had dibs on the more manageable quarry. By specializing in large game, the Neandertals may have been exploiting an ecological niche left vacant by the other carnivores. Still, the bones that have been analyzed thus far have come only from individuals who inhabited western Eurasia. It remains to be seen whether the Neandertals who lived in the Near East followed the same narrow dietary strategy.

H. sapiens, in contrast, hunted both small and large animals, and their sophisticated projectile weaponry enabled them to maintain a safe distance from their prey during a kill. Ultimately their low-risk strategy, with its dependable returns, contributed to their dominance over the Neandertals.

Although the Neandertals are, in some ways, as controversial today as they were when the remains from Feldhofer cave came to light more than 150 years ago, the arguments have evolved considerably. Clearly these hominids were not the degenerate brutes scientists once imagined them to be, and they must have been incredibly resourceful to survive for as long as they did under those punishing Ice Age conditions. Yet at the same time they were different from early modern humans, both in their appearance and in their behavior. Indeed, I believe that Neandertals and moderns were so distinct from one another in their physical appearance, hunting behavior, language, dress, customs, and so on that they would not have interbred. Humans today participate in weird sexual encounters and it is possible, but not easy to prove, that a chance mating between these two creatures did occur. But being two separate species, *H. neanderthalensis* and *H. sapiens,* fertilization probably did not occur. If it did, the fetus was aborted or the offspring was infertile.

In my opinion Neandertals are a classic example of speciation due to geographic isolation. Their ancestors became isolated in southwestern Europe and evolved for hundreds of thousands of years totally separately from *H. sapiens* living in Africa. Their long geographic separation also prevented them from exchanging genes—a prerequisite for speciation. I'll wager that when Neandertals and *H. sapiens* saw each other for the first time, mating was the last thing on their minds. More likely, when *H. sapiens* first exercised their

superb hunting abilities and outcompeted Neandertals, the Neandertals avoided any further contact.

We find it hard to envision living alongside another hominid
because we are alone today. We are the sole species of hominid on
the planet and it seems strange that there could have been, as there
was, another species of human. Also, I think that some people feel as
if we are responsible for the demise of Neandertals, because our
ancestors outmatched them and forced them into peripheral iso-
lates in the Iberian peninsula and Italy, where they hung on until
their population size dwindled and they ultimately went extinct.
When it became apparent that the sequencing of the Neandertal
genome was progressing well, some people mused that we have a
moral responsibility to bring *H. neanderthalensis* back into existence
through selective breeding, if that were even possible. It is perhaps
better to realize how fascinating Neandertals are in their own right
and how each new finding about them also brings us closer to finally
understanding what makes us modern humans unique.

The Rise of *Homo sapiens*

There are more than 6 billion people alive today, present on every continent on the planet. Some are tall, some are short; some are dark skinned, others are fair. All are members of the same species, *Homo sapiens.* But where did we come from? And when did the first of our kind arise?

A century ago, scientists such as Marcellin Boule (the French paleontologist who reconstructed the Neandertals as ogres) and English anatomist Sir Arthur Keith believed that modern humans had ancient roots. Their "pre-*sapiens*" theory held that moderns arose before the Neandertals, thus the brutish Neandertals could not have given rise to moderns. Boule's student Henri Vallois took up the torch of the pre-*sapiens* idea and carried it into the 1950s. But the scenario was eventually discredited, in part because the Neandertal fossils on which proponents based the theory turned out to be older than previously believed, and because certain modern human fossils turned out to be younger than originally thought.

Today most paleoanthropologists subscribe to one of two theories of modern human origins. One, the so-called Out of Africa model, holds that *H. sapiens* arose relatively recently in Africa and subsequently fanned out across the globe, replacing archaic hominids such as the Neandertals as it went. (This is sometimes referred to as Out of Africa II, to distinguish it from the first African exodus, which occurred far earlier in humanity's past, 1.8 million years ago.) The other, known as the multiregional evolution hypothesis, posits that modern humans emerged from archaic populations across the Old World. Franz Weidenreich, the noted German anatomist best remembered for his rigorous work on the *H. erectus* fossils popularly known as Peking Man, first proposed a polycentric scenario for the origins of modern humans in the 1930s. He suggested that although there was some gene flow between evolving populations in Europe, Asia, Australia, and Africa, he believed they evolved fairly independently into *H. sapiens.* Carleton Coon, an American anthropologist at the University of Pennsylvania, picked up Weidenreich's cause in the 1950s and suggested that distinct "racial" lines could be traced back to *H. erectus* times. But in the liberal climate of the 1960s his work found few supporters and in fact his book *The Origin of Races* presented little fossil evidence to bolster such a view. Alan Thorne of the Australian National University and Milford Wolpoff of the University of Michigan are the leading proponents of the multiregional model that replaced Coon's ideas. Since the early 1980s they have championed the view that through a succession of fossils dating back into the Pleistocene, distinct evolutionary lineages culminating in modern humans can be recognized in Europe, Africa, eastern Asia, and Australasia. Unlike Coon, however, multiregionalists believe that these evolving populations were always linked to one another through gene flow.

Over the past two decades, the Out of Africa hypothesis has gained considerable momentum from analyses of the DNA of both living people and Neandertals. Much of this genetic work has focused on a particular kind of DNA that comes from the cell's energy-producing organelles, or mitochondria. Because mitochondria are inherited from the mother alone, studies of mitochondrial DNA (mtDNA) can reveal an individual's maternal ancestry. In a landmark paper published in *Nature* in 1987, American geneticists Rebecca Cann, Mark Stoneking, and Allan Wilson concluded from their study of the mtDNA of nearly 150 people from five different populations that all of their mtDNA lineages could be traced back to a woman who lived in Africa some 200,000 years ago, a mitochondrial Eve. Since then, numerous analyses of mtDNA and nuclear DNA have converged on more or less the same recent African origin for modern humans. The latest studies indicate that *H. sapiens* originated in East Africa, between 200,000 and 150,000 years ago, and eventually made its first foray off the continent around 40,000 to 50,000 years ago. Yet critics have charged that the fossil support for the theory is much murkier. Among their complaints was the observation that if anatomically modern humans got their start in Africa, then that's where the earliest modern-looking fossils should come from. But a gap in the African fossil record between 500,000 and 100,000 years ago, when the transition to modernity is believed to have occurred, long prevented researchers from testing that prediction.

Recent discoveries have bridged that gap. The first came in 2003, when Berkeley's Tim White and his colleagues announced that they had unearthed three 160,000-year-old *H. sapiens* skulls from a site called Herto in Ethiopia's Middle Awash, the same general region about 30 kilometers south of Aramis where White found *Ardipithecus ramidus.* Two of the skulls belonged to adult males; the third is that of a six-year-old child. The more complete of the two adult male crania is large and robustly built with a strikingly tall, broad face and a heavy brow over each eye socket. Its long, high, globular braincase housed a 1,450-cubic-centimeter brain, somewhat larger than the modern human average of 1,350 cubic centimeters. The other two skulls are especially interesting in exhibiting stone tool cut marks that indicate that these individuals had been defleshed when the bone was still fresh. Extensive cut marks on the child's cranium are visible in deep nooks and crannies on the base of the specimen, which is broken away, but it is unclear whether the brain was cannibalized. Even more interesting is the observation that the cranium of the child shows polish in certain areas, which might reflect extensive handling. A 500,000-year-old skull from another Middle Awash site, called Bodo, also exhibits cut marks. But the meaning of what may have been some sort of mortuary practice remains unknown.

The hominid fossils at Herto are associated with stone tools and butchered animal remains, particularly of a hippopotamus. The tools represent a phase of stone tool technology that is intermediate between more ancient Acheulean industries that have a very high percentage of hand axes and later African Middle Stone Age industries in which flake tools are much more common. Whether the Herto people were actually hunting and killing animals is uncertain, but the hippo skull has a deep cut in the top of its skull, which is suggestive.

Following extensive comparison with some three thousand modern *H. sapiens* skulls, White's group determined that the adults are definitely more robustly built than moderns, with very rugged braincases. For this reason the team decided to distinguish them as a new subspecies, *Homo sapiens idaltu*. *Idaltu* was chosen from the Afar word meaning "elder." Interestingly, the archaic features that characterize the Herto hominids are intermediate between the very heavily built older crania like Bodo and more recent *H. sapiens* skulls from places like Qafzeh in Israel. Until the Herto finds, the oldest reliably dated examples of *H. sapiens* came from Qafzeh and nearby Skhul, which date to roughly 115,000 years ago. When the best-preserved cranium from Herto is pictured adjacent to the Qafzeh specimen known as skull 7, the African specimen foreshadows the slightly more modern look of skulls from the Levant, adding support to the widespread idea that *H. sapiens* emerged first in Africa and spread through the Middle East to Eurasia.

When the Herto finds were first published in 2003, we didn't realize that even more ancient evidence for *H. sapiens* had surfaced some thirty-six years previously, also in Ethiopia. In 1967, as part of the international effort to explore the Omo River region in southwest Ethiopia for fossil hominid remains, a Kenyan team under the direction of Richard Leakey recovered hominid crania from an area called Kibish. The Omo I skull had a large braincase, a squat face, a tall forehead, and a jutting chin, among other modern features, and was thus widely heralded as a member of *H. sapiens.* Omo II, on the other hand, appeared more archaic, with its pronounced muscle markings and a somewhat receding forehead. Yet it allied itself with *H. sapiens* in having long, arched parietal bones on the side of the cranium and a cranial capacity of 1,435 cubic centimeters, well within the range of modern humans.

Leakey's team had dated the Omo specimens using a relatively imprecise method called uranium-thorium that yielded a questionable age of 130,000 years for the Kibish hominids. In February 2005 a team led by John Fleagle of Stony Brook University went in search of the precise place and geological level from where the specimens had been collected. Renowned Rift Valley geologist Frank Brown of the University of Utah, who had worked at Kibish in 1967, accompanied the expedition. With the aid of film footage taken by National Geographic at the time of the initial find, the team was able to relocate the precise spot from which the skulls came, demonstrating how vital it is to accurately record the places where fossils are discovered.

Frank was elated—not only had the site been relocated but he spotted volcanic ash horizons that contained mineral crystals suitable for the much more accurate argon-argon dating method. Dating of the ash layer immediately above the level that had yielded the hominids revealed an age of a whopping 195,000 plus or minus 5,000 years. This made the Kibish hominids, which had resided in the National Museum of Ethiopia for decades, the oldest definitive

evidence in the world for *H. sapiens*. The team was further rewarded for their efforts with the recovery of a fragment of hominid femur shaft that attached perfectly to the bottom end of a femur fragment originally collected in 1967.

The Herto and Omo dates fit well with the molecular biologists' estimates of when and where our species originated, and they also bolster the Out of Africa hypothesis. To many researchers, the presence of modern humans in Africa while archaic humans such as the Neandertals were still developing their distinctive traits makes it highly unlikely that Neandertals were among the ancestors of our species, which is what the multiregional evolution theory predicts.

Wolpoff and other multiregionalists, however, note that although it seems increasingly likely that anatomically modern humans first appeared in Africa, these hominids probably interbred with the archaic forms they encountered when they spread out from their natal land. If so, then Neandertals and other archaic humans may have contributed to the modern human gene pool, as the multiregionalists have suggested. Studies of Neandertal DNA have yet to reveal contributions on their part, however.

Humans who looked like us had evolved by at least 195,000 years ago, as the Kibish remains demonstrate. But when, paleoanthropologists have wondered, did *H. sapiens* begin thinking like us? It's hard enough sometimes to tell what the person seated across from you is thinking, never mind individuals who lived many millennia ago. But the material culture that ancient humans left behind does give us some insights into their mind-set.

The conventional wisdom of the past twenty years has been that our species was stuck in a sort of mental rut until about 40,000 years ago, producing the same kinds of relatively simple stone tools that hominids had been making for tens of thousands of years and little

more. After that date, however, our forebears suddenly began engaging in all sorts of sophisticated practices: manufacturing advanced weaponry, forming long-distance trade networks, and creating art and music. This "big bang" theory, as it is sometimes referred to, was based primarily on the archaeological record of Ice Age Europeans, which is divided into the Middle Paleolithic (dating to before 40,000 years ago) and the Upper Paleolithic (after 40,000 years ago).

As it happens, it is during this transition from the simple Middle Paleolithic traditions to the advanced practices of the Upper Paleolithic that anatomically modern humans began establishing a foothold in Europe, which previously had been the stomping grounds of Neandertals alone. Although no human remains have turned up at the earliest Upper Paleolithic sites, researchers have traditionally assumed that the artifacts at these sites are the handiwork of moderns, not Neandertals.

More recently, Stanford University archaeologist Richard Klein has argued that the European big bang was an outgrowth of a behavioral revolution that began around 50,000 years ago in Africa, where the comparable culture periods are called the Middle and Later Stone Age. By this time in Africa, *H. sapiens* was the only hominid around, so Klein believes that the behavioral shift stemmed not from encounters with other hominid species but from a genetic mutation that conferred an ability to think symbolically. This great leap in cognitive prowess, he argues, may have been what allowed *H. sapiens* to expand out of Africa and conquer the rest of the globe. Klein draws on evidence from a site in central Kenya called Enkapune Ya Muto ("the twilight cave") to bolster his case. There, in levels dating to around 43,000 years ago, Stanley H. Ambrose of the University of Illinois and his team have found sophisticated stone tools and small disk-shaped beads carved from ostrich eggshell. The !Kung San hunter-gatherers of Botswana exchange necklaces of very similar beads to this day. Ambrose suspects the ancient residents of

Enkapune Ya Muto made these beads for the same reason that the !Kung San do today: to foster good relationships with other groups in case times get tough.

In 2002 ancient-DNA expert Svante Pääbo of the Max Planck Institute for Evolutionary Anthropology and his colleagues reported that a modified version of a gene related to language known as *FOXP2* had swept through the human population at some point in the last 200,000 years. Klein seized on this finding as evidence for this theory. But in 2007 Pääbo and his colleagues announced that new work had pushed back the origin of the *FOXP2* variant to before 350,000 years ago. As such, it no longer supported Klein's ideas. Klein, however, stands by his hypothesis, even though many scholars dismiss it on the basis that it is impossible to detect a genetic mutation with archaeological or fossil remains.

But for years experts have known of a handful of finds that are advanced for their age, and that therefore run counter to these big bang models. Among these exceptional remains are a set of three 400,000-year-old throwing spears from Schöningen, in Germany; two 100,000-year-old notched bone fragments from Klasies River Mouth Cave in South Africa; a 60,000-year-old flint incised with concentric arcs from Quneitra in Israel; a carved and polished section of mammoth ivory from Tata in Hungary that is 50,000 to 100,000 years old; and a putative figurine of a woman etched into a plum-size pebble of volcanic tuff 233,000 years ago from the site of Berekhat Ram in Israel.

More recently, at several sites in Katanda in the Democratic Republic of the Congo, Alison Brooks of George Washington University and John Yellen of the National Science Foundation have found elaborate bone harpoons that they believe are at least 80,000 years old. These implements are comparable to harpoons from Europe that are 25,000 years old, both in terms of the sophistication of their design and the selection of material: The use of bone and ivory in the manufacture of tools was thought to be a Later Stone Age/Upper Paleolithic invention. Moreover, the remains of giant Nile catfish turned up alongside some of the Katanda harpoons, hinting that early humans were keeping track of when the fish were spawning. This is significant because researchers have traditionally maintained that only later humans mapped seasonal resources.

Other evidence raises the question of whether behavioral modernity predates the origin of *H. sapiens* and even Neandertals. At a site near Kenya's Lake Baringo, Sally McBrearty of the University of Connecticut and her team have excavated stone blades—previously considered a hallmark of the Upper Paleolithic—in excess of 510,000 years old. And a nearby locality has yielded huge quantities of red ocher—and grindstones used to process it into a powder—dating to at least 285,000 years ago. McBrearty believes the Middle Stone Age inhabitants of Baringo were using the ocher for symbolic purposes, possibly to decorate their bodies, just as many humans do today. And at Mumba Rock Shelter in Tanzania excavators have unearthed 130,000-year-old flakes crafted from obsidian, the source of which is a volcanic flow nearly 200 miles away. These obsidian implements strongly suggest that their makers traded with other groups for the exotic raw material.

On the basis of these and other examples of precocious modernity, Brooks and McBrearty have argued that the dawning of modern human behavior was not a revolution, as it has so often been described, but rather an evolution. In a paper published in the *Journal of Human Evolution* in 2000, they made the case that many of the elements of modern behavior said to have originated more or less simultaneously between 50,000 and 40,000 years ago are evident at much older Middle Stone Age sites. And rather than appearing simultaneously, they emerged piecemeal, at locales across a wide range of time and space.

Yet researchers have often dismissed such examples of allegedly early sophistication on the basis of uncertainties surrounding their age and significance. That tide may be turning, however, in light of a growing body of evidence in Africa that our ancestors began acting modern long before the Later Stone Age.

The discovery of an extraordinary collection of Middle Stone Age remains from a seaside cave called Blombos, located some 200 kilometers east of Cape Town, South Africa, has significantly strengthened that notion. There archaeologists led by Christopher Henshilwood of the University of Bergen in Norway have uncovered an impressive assemblage of sophisticated tools in sediments dated to 75,000 years ago and earlier. Henshilwood's team has unearthed dozens of bone tools, including delicate awls and projectile points. These implements are remarkable because bone tools dating to more than 25,000 years ago are rarely found at African sites and have always been considered to be an innovation of *H. sapiens* that first appeared in Upper Paleolithic Europe. At Blombos, the toolmakers made a series of deliberate choices in the fashioning of these items. The first order of business was selecting the piece of bone they wanted to work. They then shaped the bones into artifacts using a variety of techniques, including whittling with a stone flake, polishing with fine-grained sediment, and burning (presumably to harden the bone). Henshilwood and his colleagues note that these tools show traces of extensive use and suggest that the Blombos people may have employed them for hide working, for the processing of plant materials, and for projectile weapons. Far from being a one-off experiment, these bone tools were clearly a central part of the toolkit.

The most startling artifacts to come from Blombos, however, are the ones that attest to symbolic thinking among the cave's long-ago occupants. The same 75,000-year-old levels that contained the bone and stone implements have also yielded engraved pieces of bone and ocher, and dozens of tiny shell beads the size of corn kernels that appear to have been strung together and worn as necklaces or bracelets.

If the work that went into producing them is any indication, these items were very important to their makers. In his analyses of two of the engraved ochers, Francesco d'Errico of the University of Bordeaux has determined that the reddish rocks were ground on one side to produce a facet that was then etched with a stone point. Bead production, too, was labor-intensive. Henshilwood believes the shells were collected from estuaries located 12 miles from the cave. And experimental reconstruction of the technique employed to perforate them suggests that the Blombos jewelers used bone points to pierce the lip of the shell from the inside out—a method that commonly broke the shells when members of Henshilwood's team attempted it.

When the Blombos discoveries came to light, some researchers viewed them as incontrovertible proof of cognitive sophistication in the Middle Stone Age. But others were unpersuaded, noting that the artifacts from Blombos were the exception, not the rule. To that end, material from a group of sites in nearby Mossel Bay, South Africa, is likely to erase any lingering doubts.

It's a clear morning in Mossel Bay, a picturesque town along the famed Garden Route of South Africa's Western Cape province. The Indian Ocean glitters with the early light while the Outeniqua Mountains hold court in the distance. Tourists flock here for the balmy weather, the beautiful beaches, and the delicious mussels for which the bay is named. I'm here for a different reason: I've come to visit a series of spectacular Middle and Later Stone Age archaeological sites that my IHO colleague Curtis Marean and his team are excavating. So rich are the deposits here that observers have called the area "Les Eyzies of Africa," after a town in southwestern France that is home to one of the richest concentrations of Upper Paleolithic remains in Europe.

Tall and athletic, with a knack for innovation, Curtis left graduate school in 1990 determined to make a significant mark on

African archaeology. He has distinguished himself by developing excavation and recording methods that have set new standards for the field, and his methodology for analyzing animal remains from archaeological excavation has revealed more about Paleolithic hunters than many thought possible. Curtis's outside-the-box thinking permeates his entire research agenda, from documenting paleo-climate change in southern Africa and understanding the roots of modern human behavior to viewing Neandertals from an African perspective rather than a Eurasian one. I was thrilled when he agreed to join the IHO research team in 2001 and relocated to Arizona State University from Stony Brook University.

With extensive field experience at sites in East and South Africa, including Blombos, Curtis had a gut feeling that whereas the fossil record for early hominids was excellent in eastern Africa, the chances of documenting the beginnings of modern human behavior were far superior along the South African coastline. His quest for cave sites rich in archaeological deposits led him to the area of Mossel Bay. There, at a place called Pinnacle Point, he located a series of caves that archaeologists had left unexcavated because of their location on a very steep cliff face. It was important that he find sites that were at least 15 meters above the current sea level, because elevated sea levels 123,000 years ago would have washed out all the archaeological remains deposited in them. Curtis was in luck: The caves at Pinnacle Point are situated well above the ocean's reach. Thus in 2000 he chose to begin excavating one of these caves, known simply as 13B.

Curtis and I stood on the edge of the cliffs at Pinnacle Point looking out over the pristine sea as right whales and dolphins cruised by. It sure beat working in the dry, dusty, hot deserts of Ethiopia. I could see the waves crashing some 20 meters below the dark cave opening of 13B. The only way to the cave when Curtis first arrived here was by rappelling down a sheer rock face. He has since rigged a 180-step steel staircase anchored into the treacherous cliff face that leads into the cave by way of a small bridge.

As we descended the staircase, Curtis explained that not only was this area archaeologically rich beyond belief, but the caves preserved a wonderful record of paleoclimate and paleoenvironment that would provide a context for human evolution here in the Cape. It is no wonder he was awarded the largest archaeological grant in the history of the National Science Foundation, $2.5 million for his multidisciplinary project at Pinnacle Point that integrates the expertise of more than thirty international scientists from a wide range of specialties.

When we finally entered the cave, it was a hive of activity. Some of Curtis's students were meticulously excavating its ancient layers by the light of their miner's lamps; others were recording each and every fragment of bone and stone recovered using laser-guided survey equipment. Looking back out of the cave opening I could see the ocean and hear the waves crashing against the rocky shore below: what a spot.

Difficult though it was to make the caves at Pinnacle Point accessible, the effort has paid off in spades. In October 2007 Curtis and his team published a paper in *Nature* highlighting some of their discoveries at 13B. Gratifyingly, Curtis's instincts turned out to be spot-on: The cave has yielded the earliest example of systematic exploitation of marine resources. In levels dated to 164,000 years ago the excavators found multiple species of shellfish, mostly the same brown mussels that people there eat today. They also recovered a large whale barnacle, suggesting that the Middle Paleolithic occupants of 13B scavenged the skin and blubber of a beached whale. Curtis believes that these prehistoric foragers were collecting food from the rocky intertidal zone, which would have required surprisingly sophisticated planning. "People often think that shellfish were an easy food resource to exploit" because they don't run away, he observes. But in rocky intertidal zones, shellfish like brown mussels are covered by water most of the time. Entering these areas can be quite dangerous—it's easy to get knocked off the rocks and killed by an incoming swell. The only time the shellfish are exposed and safe to collect, Curtis explains, is during low spring tides, when the gravitational pull of the moon is strongest. (These tides are so-named not for the season, but for the way in which they spring back and forth.) Because the tides are tied to the lunar cycle, he believes the residents of 13B were scheduling their trips to the shore using a lunar calendar. "Tides advance at fifty-minute increments a day. This is why if you're a fisherman, you have a tide chart," Curtis remarks. "Imagine trying to track a tidal situation without a tide chart."

Based on the division of labor seen in modern hunter-gatherers, it seems likely that women were the ones gathering the shellfish in Mossel Bay 164,000 years ago. And because the team has found hearths at 13B with burnt shell in them, Curtis believes these early *H. sapiens* were cooking the mussels and other bivalves. "If you don't have pots and pans, cooking shellfish isn't easy," he comments. But after researching how people prepare brown mussels today and conducting numerous experiments, he concluded that they may have pressed them into the sand and poured hot ashes over them. "This way the shell stays closed and the juices are retained," he says. However they were prepared, these ancient shellfish push back the earliest evidence for human use of marine resources by nearly 40,000 years.

That isn't the only advanced behavior evidenced at 13B, how-ever. Workers also unearthed a number of bladelets—long thin flakes too small to use in one's hand—that probably would have been mounted in bone or wood and used as projectiles. Archaeolo-gists consider such composite tools, as they are known, to be fairly complex, because they require more planning than do handheld stone tools. The Pinnacle Point bladelets more than double the age of this technology.

Additionally, the team has uncovered fifty-seven pieces of red ocher at the site, which Curtis interprets as indicative of symbolic behavior. Of these, the inhabitants seem to have preferentially used the reddest pieces, grinding their surfaces to produce powder. He believes they were mixing this powder with other things—animal fat, perhaps—to create a sort of paint. The choice of red seems to have been intentional. Many different colors of ocher are found in the area, but inside the cave, the vast majority of the ocher is bright red. Why did these early modern humans favor the scarlet variety? We can't know for sure. But "red is often associated with blood, menstruation," Curtis reflects. "There's a fertility implication there."

It may not be coincidental that these early coastal people also thought symbolically. When humans start to rely on seafood, group sizes get bigger and mobility decreases, Curtis explains. "When that happens, a lot of social complexities arise" and people have to learn to live with many more people. The early use of pigment at 13B "could tie into a need for social cooperation," he notes. After all, "people use things like jewelry, body decoration, and so on to communicate messages to people."

The modern behavior of the 13B people 164,000 years ago was no flash in the pan. Younger materials from Pinnacle Point are similarly advanced. At another site the team discovered a 100,000-year-old crayon made of ocher the consistency of lipstick, and excavations at a rock shelter known as PP5-6 have revealed evidence of a remarkably advanced stone tool tradition spanning the time from 93,000 to 45,000 years ago. In a presentation given at the annual meeting of the Paleoanthropology Society in 2008, Curtis's colleague Kyle Brown reported that the stone used in the manufacture of most of the bifacial tools at the site appeared to have been heat-treated before the ancient knappers shaped it. The researchers noticed that they couldn't find any local sources of the material—a stone called silcrete—that matched the color and luster of the silcrete implements themselves. Brown noticed, however, that the knapped silcrete resembled that used in certain Navajo artifacts that he knew to be heat-treated. In subsequent experiments, the team found that heating the silcrete gave it the color and luster of the silcrete used in the ancient tools. The researchers also observed that the heated samples were significantly harder than untreated silcrete, and much easier to knap. Furthermore, twelve of the thirteen archaeological samples tested were found to have been heated prior to tool manufacture. This earliest known instance of heat treatment of stone is a testament to the intellect of our prehistoric forefathers.

Geneticists believe that sometime around 140,000 years ago, the founding population of modern humans underwent a catastrophic event that slashed their numbers from around 12,800 breeding individuals to a mere 600. Those 600 people gave rise to the modern humans who would one day leave Africa and colonize the rest of the world. In other words, they were the ancestors of every human alive today. What was it that nearly drove our species to extinction so soon after it first evolved? Our best guess is that climate may have been to blame. We know that between 195,000 and 130,000 years ago, glacial conditions gripped the planet, leaving most of Africa inhospitably cool and dry. Only a handful of places on the continent could have supported human populations during that glacial period, Curtis says. And he suspected that the southern coast of South Africa might be one of them, thanks to its warming ocean currents and associated abundance of shellfish as well as its fynbos vegetation, which has the highest diversity of endemic plants of any floral kingdom. Cape Town's famed Table Mountain, for example, has more species of plants than exist in all of the British Isles. Particularly important for early humans who lived there would have been the large diversity of nutritious tubers, which are high in carbohydrates and thus a perfect complement to the shellfish.

With all of this success behind him Curtis is articulating an even more ambitious plan for his work on the Cape that will dominate his research there for years to come. Currently he is crafting strategically targeted fieldwork encompassing a transdisciplinary approach that will tell him whether the progenitor population of all humankind indeed inhabited this part of Africa. The southern Cape may well prove to be the Garden of Eden.

CHAPTER 16

Unsolved Mysteries

The question of questions for mankind—the problem which underlies all others, and is more deeply interesting than any other—is the ascertainment of the place which Man occupies in nature and of his relations to the universe of things.

—THOMAS HENRY HUXLEY, *EVIDENCE AS TO MAN'S PLACE IN NATURE*, 1863

Some 3.2 million years after she died Lucy has become an international celebrity, and Ethiopia's prime minister, Meles Zenawi, decided that her skeleton should tour American museums as a goodwill ambassador. The remarkable and distinguished history of Ethiopia is not well known in the Western world and the prime minister hoped Lucy would change all that. The exhibit, entitled "Lucy's Legacy: The Hidden Treasures of Ethiopia," premiered at the Houston Museum of Natural Science in August 2007, where it received some 175,000 visitors. (That the exhibit and this book share the same title is a coincidence, however—the projects are independent of each other.)

In November 2007 His Excellency Girma Wolde-Giorgis, the president of Ethiopia, invited me to attend a gala dinner at the museum, where a significant monetary contribution was presented to the Ministry of Culture and Tourism in support of Ethiopian paleoanthropology. When the president and I were led through a dark corridor to the display case that held Lucy, my heart began to beat a little faster. The last time she had been in the United States was twenty-eight years ago, when I was working on the detailed description of her remains at the Cleveland Museum of Natural History. Standing before a bulletproof glass case we gazed in silence as the bright overhead lights illuminated her skeleton, splendorous on a black velvet cloth. After a few moments of awe I explained to His Excellency how the discovery of Lucy had launched a highly successful era of fossil exploration that now distinguishes Ethiopia as the single most important country in all of Africa for documenting human evolution over roughly the last 6 million years.

Ethiopia's leadership role in African paleoanthropology was highlighted at an international conference that took place on January 12–15, 2008, in Addis Ababa. The gathering was opened by Ambassador Mahmoud Dirir of the Ministry of Culture and Tourism, which organized the meeting in conjunction with the Ethiopian Millennium Festival National Council Secretariat. How fitting that Ethiopia would celebrate its Millennium (Ethiopia still uses the Egyptian Coptic calendar, which is eight years behind our own) with a conference dedicated to the origins of humankind.

It was deeply rewarding to see so many young Ethiopian scholars presenting results of their research. When I first began work in Ethiopia there was not a single Ethiopian scientist conducting human origins research, but now more than a dozen researchers are crafting the future course of paleoanthropology and related disciplines in their native country.

Ethiopia's commitment to its role as the Cradle of Humanity is no more evident than in the modern, six-story paleoanthropology building that the government has built in Addis Ababa. The building contains private offices for visiting scholars and study and storage space for millions of specimens; it will be the focus for more than twenty-five field projects already under way in Ethiopia. Research here will focus on conservation, archaeology, paleontology, and paleoanthropology, as well as art and history. No other country in

Africa is currently pledging such a level of resources to the study of paleoanthropology. There are even plans to construct a major museum in Addis Ababa solely dedicated to human origins.

Each spring I teach a course called Human Origins to about a hundred students at Arizona State University. The anthropology majors regularly show up in my office to debate one or another topic, but mostly to inquire about the wisdom of choosing a career path in paleoanthropology. The most frequent refrain I hear is, "Well, it seems there is nothing left to do in paleoanthropology." They are surprised when I tell them that this is a great time to be in the field because more and more fossils are being found and the breadth and depth of questions being asked about them is really limitless.

Admittedly the study of human origins is a relatively new science. Only since the late 1960s has the full extent of what is now a truly international multidisciplinary endeavor been scoring huge dividends. No doubt paleoanthropology has a long road ahead before all the deepest questions about our ancestry are answered, but we've come a long way since David Pilbeam of Harvard University wrote in a 1978 review of Richard Leakey's book *Origins* "that perhaps generations of students of human evolution, including myself, have been flailing about in the dark; that our data base is too sparse, too slippery, for it to be able to mold our theories."

In the early days of human origins research the major goal was to find fossils, as if a particular specimen would somehow be the Rosetta Stone of human evolution. As more and more finds were made it became obvious that human evolution was much more complex than anyone had envisaged. Each precious fossil find was only a part of a much larger puzzle. We began to ponder the when, where, and why of human evolution. The field is becoming more problem oriented, and although the thrill of discovery will never fade, we are now increasingly focused on the many unanswered questions about how we became human.

One major question facing us is who the last common ancestor of hominids and the African apes was. Discoveries from Chad, Kenya, and Ethiopia, as well as the use of molecular clocks calibrated on DNA work, indicate that this ancestor lived in Africa between 6 and 10 million years ago. In 1967 Vincent Sarich, an emeritus professor at Berkeley, and his Berkeley colleague Allan Wilson (now deceased) published their study on human and primate proteins as a proxy for the degree to which humans and other primates differed. Their stunning conclusion that humans and apes separated only 5 million years ago ignited a firestorm of opposition from paleontologists, who subscribed to the idea published by David Pilbeam that the ape-human divergence occurred 15 to 28 million years ago.

Over the last forty years increasingly sophisticated methods for assessing similarities and differences between the DNA of different species have shown that humans and chimpanzees are more closely related to each other than chimps are to gorillas or gorillas to us. That confirms a branching pattern that has gorillas splitting off from a Miocene ape ancestor prior to the split between chimps and humans, which several studies suggest occurred somewhere between 4 and 8 million years ago. Gorillas may have diverged some 2 million years earlier.

While the assumption about the rates of evolution may not be without problems, the current view is that the last common ancestor to chimpanzees and humans was much closer to Sarich and Wilson's date than paleoanthropologists expected. Perhaps as the genomes of other primates such as the gorilla and bonobo become better known a more precise date of chimp-human divergence will become available.

Although humans and chimps differ remarkably in appearances and behavior, they share some 98.8 percent of their genetic identity. With the successful sequencing of the human genome and, more recently, the chimp and macaque genomes, geneticists are beginning to scan regions of those genomes for differences and even starting to identify the precise function of specific genes. The loss of the hair keratin protein gene in humans, which apparently occurred only some 240,000 years ago, indicates that our ancestors were pretty hairy until relatively recently. And the loss of some genes that are important for smell and taste in chimpanzees has also been documented, reflecting our lowered dependence on these senses. Scientists have also found that one region in the human genome associated with development of the brain's neocortex and another associated with perception and memory areas of our brains have undergone positive selection in the human lineage and not in the apes.

Still, the exact nature of the last common ancestor of chimps and humans remains a mystery, and we may never be able to identify it. As we dig deeper into the past, hominid fossils become rarer and rarer. Except for a relatively few unique features such as those related to bipedalism or loss of apelike canines, it will be quite a challenge to assess whether future fossil discoveries belong in the human lineage. The current evidence suggests that bipedalism was *the* defining feature, and we can only hope that even older fossils will contain parts of a pelvis, knee, or foot to confirm that adaptation and allow us to designate a specimen as hominid. Hopefully the earliest stem-hominid will also preserve a suite of primitive features that will permit us to tie that species directly back to a specific Miocene ape ancestor.

Another truly interesting and answerable question concerns the diversity of species of australopithecines that lived; present evidence indicates there were more than half a dozen. Perhaps the best understood is *A. afarensis,* but little is known about its anatomical uniqueness and evolutionary relationships or the lifeways of the other species. What was it about the world of australopithecines that promoted this diversification? What was the relationship between robust australopithecines in East and South Africa—are we really looking at independent evolution in each region? Did they have a common ancestor? Did they arise in response to climatic change? What was the world like 2 million years ago when perhaps several species of *Homo* and *Australopithecus* wandered the African landscape and even overlapped geographically? How did they partition the world in which they lived—like chimps and gorillas do in areas they cohabit today? As the only surviving hominid, we are accustomed to being alone, but how would we feel if there were "others," species of our own genus? How thrilling, or maybe frightening, might that be?

Perhaps one of the most exciting and least understood aspects of hominid evolution is that decisive period between 3 and 2 million years ago when our genus, *Homo,* made its first appearance. What were the circumstances leading up to this debut? A number of events in human evolution have been linked to climate change: our separation from the common ancestor with the African apes; the advent of bipedalism and the appearance of stone tools; perhaps meat eating, brain expansion, and the dawn of *Homo.* However, we are a long way from a comprehensive understanding of the interaction between climate change and human evolution.

When I was in graduate school it was universally agreed that hominids did not migrate out of Africa until perhaps some 500,000 to 700,000 years ago. Most believed that until our ancestors grew big, controlled fire, and developed sophisticated tool kits, such an exodus would have been impossible. It would have just been too daunting to leave the comfort of a tropical Africa for the challenges of temperate Eurasia. With the possibility that hominids reached Southeast Asia 1.7 million years ago and the discovery of *H. ergaster* in the Republic of Georgia, paleoanthropologists have had to reconsider those arguments. The Georgian hominids are well dated at 1.77 million years, they are accompanied by primitive Oldowan tools, and it is unlikely that they controlled fire. So, when, how, and why did they leave Africa? Were they just curious, fascinated as all humans are by exploration to find out what is around the next mountain, across the river, or beyond the forests? Were they simply increasing their territory by pursuing new food sources or following migrating herds? If hominids made it out of Africa by 1.77 million years ago, maybe earlier species also made the journey and we just have not yet found their remains. Could the diminutive Flores hominids be an isolated group of *H. habilis* that ventured afar at a time that no paleoanthropologist would have thought probable?

The Neandertal problem has come under increasing scrutiny and we have moved light years from the notion that they were dim-witted brutes hanging on by a thin evolutionary thread. They survived in Eurasia for hundreds of thousands of years under brutal, glacial conditions and really only began their decline when a more sophisticated and culturally advanced *H. sapiens* began to challenge them. What was the nature of Neandertal-modern human encounters? Was there intense conflict and aggression, or was there any exchange of experience or material goods? The exact nature of Neandertal extinction is still subject to widespread debate and there is much to learn about the time, place, and cause of their disappearance. But paleoanthropologists are now asking the right questions and getting closer to fleshing out the answers.

We have far to go in our understanding of how *H. sapiens* made it through the apparent genetic bottleneck that occurred roughly 140,000 years ago during an especially challenging period of global cooling. If the geneticists are correct in their conclusions that the founder population for all of humanity was indeed African and consisted of only perhaps hundreds or a few thousand individuals, what does that say about genetic variability for modern *sapiens*?

The renowned Russian geneticist Theodosius Dobzhansky once remarked, "All species are unique, but humans are uniquest." Picking up on this, a colleague of mine at ASU, Kim Hill, a leading authority on hunter-gatherers, put it very simply when he observed that should an extraterrestrial life form land on Earth, *H. sapiens* would attract the most excited attention of any species living on the planet because a remarkable set of abilities distinguishes us from all other organisms. Kim is convinced that the roots of our uniqueness lie in our hunter-gatherer past. He points out that for 99 percent of the time the genus *Homo* has existed, hunting and gathering has been *the* survival strategy.

Kim lists as one of our most distinctive attributes our overwhelming dependency on social learning that necessitates the successful transmission of learned behavior from one generation to the next. Other primates have varying degrees of social learning, but none as complex as ours. He also notes that we are the only primate that relies on laws, social conventions, and a moral code that defines and regulates social behavior. But perhaps the most unique behavior of all among humans is that we live in a society based on symbolism, which underlies our use of language. All animals communicate with each other, but human language facilitates the transfer of information about events that have taken place in the past, or that will be occurring in the future. We can even describe to one another events that only one of us has seen. Nonhuman primates lack these abilities and their communication is more situational. That is, it's based on what is happening around them at a particular time. In a seminar at IHO, Kim defined the exclusive nature of human language as the ability to "exchange ideas about unlimited topics that range from the concrete to the purely abstract."

The explanation for this exceptional uniqueness of ours has to lie in biology, the fossils, and the archaeological record. Our intelligence, the way we choose our mates and raise our children, our long life span, and other distinctive traits must be rooted in the past. Neandertals, too, were large-brained and probably had a life history similar to our own, so why didn't they become the dominant species on Earth? For Kim it is essential to explain why we survived and Neandertals didn't. How did our hunter-gatherer life differ from the ways of the Neandertals?

No one can predict the future, but everyone wonders just where we are going as a species and wants to know if human evolution has stopped. Well, technically every time a sperm fertilizes a human egg there is genetic recombination that produces a singular individual, different from every other *H. sapiens* that has ever lived, or will live. But the level of uniqueness is a reflection of minor differences that are far from qualifying the individual as a new species. The famed evolutionary biologist Ernst Mayr said that reproductive isolation was a prerequisite for new species to arise. Some physical or even biological barrier must prevent the exchange of genes between individuals, and if these populations remained isolated long enough new species would appear. Humans exchange genes every day all around the world and do not live in isolated reproductive groups, so the conditions for speciation are simply not present in *H. sapiens* today.

However, if humans were separated into discrete, disparate, reproductively isolated subpopulations for a very long time— because of a devastating pandemic, for instance—new species of *Homo* just might arise. Another situation in which a new species of *Homo* might form is if a mission into space returned to Earth after a few hundred thousand years of separation, during which genes could not be exchanged between those in space and those at home. How strange it would be if the returning astronauts were incapable of interbreeding with the very species that sent them into space. But how that would play out is fodder for writers of science fiction.

Evolution among humans has not stopped; in fact, according to population geneticist Henry Harpending of the University of Utah, it has accelerated. He and his colleagues analyzed DNA from four populations—the Han Chinese, the Japanese, the Yoruba tribe in West Africa, and northern Europeans—targeting 3.9 million SNPs (pronounced "snips"), which are single point mutations. After careful analysis Harpending and his coworkers concluded that 7 percent of human genes are undergoing rapid and recent evolution, challenging the notion that humans have not changed much since their dispersal from Africa some 40,000 to 50,000 years ago. Even more surprising was the conclusion that humans on different continents are becoming increasingly different due to lower gene flow between groups and the appearance of favorable mutations that enhance reproduction and survival. Some of these differences are related to cultural and ecological factors, such as selection for lighter skin in Eurasia and loss of lactase-tolerant genes in Asians and Africans. Others certainly reflect disease resistance resulting from changes in population size and structure associated with the invention of agriculture. Although the levels of change are relatively small and do not signal impending speciation in *Homo,* they do call into question the oft-cited view that human evolution should have slowed down as culture increasingly buffered humans against natural selection.

As the most egotistical of all species, we humans tend to think of ourselves as the inevitable consequence of some inherent process of evolution that preordained our existence. This has led to the single most widespread misperception of human origins, one that is commonly depicted in diagrams that portray the human evolutionary journey as a march from apeness to humanness. The image inevitably features a quadrupedal ape to the left and through a series of seemingly goal-directed evolutionary steps, the figures on the right become more upright, less hairy, flatter faced, and larger brained. (This relentless progression from ape to angel always culminates in a white European male as the Grand Finale of human evolution, probably because European males draw these figures.) The central problem with this walk through time is that it promotes the all-too-common misconception that the whole purpose of evolution was to create us.

The fossil evidence for human evolution is consistent with the idea promulgated by Darwin that the tree of life is a branching one with numerous lineages, not a single evolutionary lineage from ancient to modern. So how do we explain the preponderance of the single lineage model? If we liken our family tree to a river, like the Nile, it will become clear why so many people have that view and we will see that it is really a matter of how we perceive the evidence.

Let's get into a boat at one of the distributaries in the delta where the Nile enters the Mediterranean Sea. As we row upstream in search of the source of the Nile, we come to a fork. We choose the one that flows upstream. After all, that is in the direction of where the water has come from. Every time we reach a fork, let's choose the one that flows upstream and just ignore those that flow downstream. All the while we're making a map of our trip, which appears to us to be a straight course from the delta to the source of the Nile—the unilineal model.

But now let's get back into the boat at the source of the Nile, somewhere around Lake Victoria, and try to find our way to our original point of departure in the delta. The first thing we're going to discover is that all the forks are now flowing downstream. When a branch peters out and disappears we row back to the fork and follow the other branch; we come to another fork, and another, and another. When we reach our starting point in the delta, our map would look totally different, reflecting hundreds of branches. Experienced this way, we understand that the river branches endlessly, and is not a single, straight line from origin to culmination—the multilineal model.

If we look back in time from where we are today and choose fossils that are more and more ancient, it appears we have descended along a single lineage. Yet if we stand on the shores of the lake where Lucy died and try to follow the evolutionary path to modern humans by studying geologically younger fossils, we encounter bifurcations along the way, most of which went extinct, and we have to retrace our steps to get back on the lineage that survived. From this perspective we see how complex the human family tree actually is and realize that most branches died out. Extinction was the rule, survival the exception, as it has been ever since life began on the planet.

Another misconception generated by the idea of unilineal evolution is that ancestral species were somehow imperfect ones on their way to becoming something else. But becoming what? Do we think of ourselves as an imperfect species on the way to some goal?

No. We are who we are, just as Lucy was who she was. She wasn't half-way to anywhere, waiting for natural selection to refine her anatomi-cal and behavioral adaptations. How could her species have been on the way to *H. sapiens* when evolution has no preconceived goal? There is no intrinsic force driving evolution in any direction, and hominids, like all other organisms, were not being led by natural selection toward any predetermined goal. Natural selection can operate only on variation that is present in a species. Those charac-teristics that are unfavorable, that cause early death, are not passed on because the individuals that possess them do not survive to repro-duce. Features, presumably like bipedalism, that increase the repro-ductive fitness of a species are passed on to succeeding generations. We have always thought of the process of natural selection as "sur-vival of the fittest" when it might be more illustrative to think of it as "elimination of the unfit."

Early hominids including *A. afarensis* were not consciously or unconsciously moving toward modernity—they were totally success-ful in their own right. As is true for all species, those *A. afarensis* indi-viduals who did not survive and reproduce did not leave genes behind; those who did left their DNA imprint on descendant gener-ations. On the family tree, *A. afarensis* sits at a pivotal place between older, much more apelike creatures, such as *Ardipithecus, Sahelanthro-pus,* and *Orrorin,* and younger *Australopithecus* and *Homo.* We only know in what directions *afarensis* evolved because we have fossil remains of their descendants in the form of other australopithecines such as *A. garhi,* the robusts, and, of course, the genus *Homo.* View-ing Lucy from this perspective we see how vital her role was as a link between her ancient ancestors and younger descendants.

Lucy's species prospered over a large geographical region of Africa and survived for more than perhaps 800,000 years. This is pretty successful compared to modern humans, who have been around for less than one fourth of that time. Therefore we must try to understand each and every fossil species within their own world, not ours. As challenging as it is, we have to envision them as living, breathing, reproducing individuals just like ourselves in order to appreciate how their adaptations, anatomy, and physiology permitted them to flourish in the world they occupied.

We are a fortunate species because we have the curiosity and the ability to dig up our past and piece together our long evolutionary journey. Imagine if we were chimps interested in our past, sort of paleopongidologists. We would be endlessly frustrated because of the lack of fossil ancestors. Except for three 500,000-year-old chimpanzee teeth found in northern Kenya the geological record in Africa is devoid of any fossil apes that are directly ancestral to modern-day African apes. This lack of fossil apes is partly a reflection of the acidic soils in tropical forests that are detrimental to bone preservation. We know these African ape precursors must have been there, but none of their fossils have turned up. In the end the paleopongidologist would never really know how she became chimp.

We humans, on the other hand, now have at least the broad outlines of our evolutionary journey. The fossil evidence for human evolution gives us an interesting and provocative answer to the question "Where did we come from?" This remarkable gift allows us to better know ourselves and our place in nature.

The Discovery of Ardi

On October 2, 2009, the world met "Ardi," a largely complete female hominid skeleton from Ethiopia dated to 4.4 million years ago. In 1994 a team led by Berkeley's Tim White discovered the specimen, which forms the centerpiece of the species called *Ardipithecus ramidus* (in the Afar language, *ardi* means "ground" and *ramid* means "root"). Some 1.2 million years older than Lucy, Ardi is the most complete hominid from beyond the four-million-year mark. As such, she stands to shed considerable light on this very murky period of human evolution. Indeed, so significant are Ardi and the other *Ard. ramidus* fossils that when White and his colleagues finally unveiled the specimens last fall, the journal *Science* published a whopping eleven papers aimed at describing and interpreting them. A media blitz accompanied the scholarly publications. Newspapers, magazines, evening news programs, and blogs around the world trumpeted the finds, and just ten days after the *Science* papers appeared, the Discovery Channel aired a slick two-hour documentary on Ardi. Fifteen years in the making, it was a sensational coming-out party.

Ard. ramidus is not the oldest hominid on record—*Ardipithecus kadabba* from Ethiopia, *Orrorin tugenensis* from Kenya, and *Sahelanthropus tchadensis* from Chad are all significantly older than it is, as described in chapter 9 of this book. Nevertheless, *Ard. ramidus* is

being held up as a Rosetta stone for understanding the very beginnings of the human lineage and even the last common ancestor of
humans and chimpanzees. Some of the specifics of this argument
are on firmer footing than others, however, and although there is
no doubt that *Ard. ramidus* offers up something wonderfully new for
paleoanthropology, exactly what this species means for human
evolution is a matter of some debate.

To fully grasp the significance of Ardi and her kind, we must start, as
always, at the beginning. The day was December 17, 1992. White's
team was surveying a site called Aramis, situated in the desolate and
parched landscape of the Afar Triangle's Middle Awash region,
some sixty kilometers south of Hadar, when Gen Suwa of Tokyo University found a right upper third molar. It was merely a scrap, but it
encouraged the team to keep searching the area for hominids. Since
the discovery of that tooth, the team has harvested 109 specimens
representing 36 individuals, the Ardi skeleton being the most complete (the rest of the specimens are individual teeth, tooth chips,
and bits of skeleton).

These fossils were hard-won, particularly Ardi. So fragile were
her off-white bones when the team found her that they crumbled
when touched. Not only had the skeleton been trampled by animals,
but it was not heavily fossilized. Extracting the delicate remains from
the surrounding clay required that the excavators first harden them
with a consolidant and then employ a variety of implements ranging
from porcupine quills to dental picks, in order to gently free the
bones from the clay matrix.

Most of this fossil-preparation work was conducted back at the
museum in Addis Ababa, where researchers removed the clay layer
by layer under a microscope in a laborious process they described as
"microsurgery." Considering the poor preservation of the skeleton,
it is truly astonishing that White and his team recovered so much of
Ardi. Had they not come across the bone scatter at Aramis in 1992, a

torrential desert rain could have very well washed Ardi away in a mudflow, never to be seen again.

The sedimentary layer from which Ardi was recovered appears to have been laid down in relatively calm water, and the fine clay encasing the bones also preserved fossil wood, casts of plant roots, and even seeds. Furthermore, the team recovered from this same geological horizon an additional six thousand specimens of vertebrate animals ranging in size from shrews to elephants.

Serendipitously, this fossil-bearing layer is sandwiched between two volcanic ashes dated almost identically at 4.4 million years ago, which implies that it accumulated in only a few hundred or a few thousand years. That all of the remains are confined to such a narrow slice of time increases the likelihood that these organisms are actually part of the same ecological setting, as opposed to having been mixed together by moving water and other forces of nature.

Based on analyses of these plant and animal remains, it appears that *Ard. ramidus* lived in wooded areas. The majority of the twenty-nine species of birds found at Aramis resemble living terrestrial species, not aquatic forms. The bats, shrews, rodents, and hares from the site also attest to a woodland setting rather than an open one. Among the antelopes, browsing forms like the kudu dominate the collection, whereas open grassland grazing forms are generally absent or occur only in much lower frequencies. The very high number of leaf-eating monkeys, colobines, further evinces a paleo-environment that was heavily treed.

Less straightforward than reconstructing the habitat of *Ard. ramidus* was reconstructing Ardi herself. Her crushed and highly distorted skull in particular presented a formidable challenge to the discovery team. Even in the lab it was impossible to properly clean the soft edges of various fragments without worrying about losing information vital for refitting individual pieces of the skull. The team decided to take a two-pronged strategy to putting Ardi's head back together again, molding the pieces for physical reconstruction and employing thousands of micro-computed tomography scans to represent the distorted skull and reassemble it virtually on a computer.

The team's reconstruction of the skull seems reasonable, judging from the photos that appeared in the *Science* papers, although I would prefer to examine the remains in person to make my assessment. (In fact, I was in Ethiopia when the team published their findings, and obtained permission from the museum authorities in Addis Ababa to see the fossils. But Berhane Asfaw, co-leader of the Middle Awash team, declined to show them to me.) The skull exhibits a jutting face, a low sloping forehead, and thin brow ridges. And the jaws are lightly built and set in a small face, reminiscent of much older Miocene apes. Ardi also had a small brain—around 300 cubic centimeters (cc), plus or minus 10 cc—which is well within the range of chimpanzees and bonobos.

Ardi's teeth hold clues to her diet. The back teeth, molars, and premolars used in crushing and grinding food are low-cusped and not enlarged as in *Australopithecus*. And the teeth exhibit thin dental enamel, albeit not as thin as in chimps. Combining these details with observations of enamel wear, it looks as though the diet of *Ard. ramidus* did not require a great deal of chewing. *Ard. ramidus* appears to have lived on a varied diet based on foods found on the ground and in the trees, but unlike later *Australopithecus,* whose teeth bore thick enamel, it did not eat tougher foods like ripe fruit or fibrous plants.

Although many aspects of Ardi's skull are primitive, several hint at hominid status. One such feature is the reduced size of the canine teeth, which is observable not only in Ardi but in other *Ard. ramidus* specimens, male and female. Typically, male monkeys and apes have large, daggerlike canines that overlap with and thus sharpen against the lower first premolar. In *ramidus* this sharpening function has been lost, as it has been in all hominids. And the length of the canine in both sexes is dramatically reduced compared to the ape condition. The base of Ardi's skull is also short from front to back—another hominid trait. And the hole at the bottom of the skull through which the spinal cord passes is forwardly placed, as it is in hominids, as opposed to the rearward foramen that magnum apes have.

Ardi also has a mosaic of apelike and humanlike features in her pelvis. Badly crushed when the team uncovered it, the pelvis had to be pieced together virtually because the fragments were too friable to be separated from the clay matrix. The reconstruction shows a humanlike rotation of the blade of the pelvis, which would have brought the gluteal muscles into a position that stabilized the hip during upright walking. Yet the lower portion of the pelvis displays an apelike elongation that would have positioned the hamstring muscles for powerful climbing.

Aside from the skull and the pelvis, though, the rest of Ardi's skeleton lacks much in the way of hominid traits. She has incredibly long arms and large hands that reach below the knee. In fact, the relative length of her upper limbs calls to mind that of modern-day monkeys that spend a lot of time traveling along tree branches on all fours. (Standing around 1.2 meters high and weighing about 50 kilograms, Ardi was much larger than any living monkey. For that matter, she was also considerably larger than Lucy, who was only a meter tall and weighed less than 30 kilograms.) Ardi's hand bones are also indicative of this above-branch locomotion, as it is known, with long fingers and a powerful, grasping thumb. All modern apes, in contrast, have hands built for hanging from branches rather than balancing atop them, and the African apes—our closest living relatives—exhibit specializations for knuckle-walking when traveling on the ground. Detailed study of Ardi's hands revealed no sign that she made a habit of hanging or knuckle-walking, however.

Ardi's foot is likewise perplexing. Perhaps the most startling aspect of Ardi's entire skeleton is her big toe, which is huge and sticks out to the side. In humans the big toe is aligned with the other toes and helps to propel us forward during upright walking. Other primates, which have divergent big toes, use them to grasp branches during climbing. Bruce Latimer, a member of White's team, once wrote that an adducted large toe (that is, a big toe that lines up with the rest of the toes) is an essential element of bipedalism. Yet Ardi's great toe is even more divergent than that of a chimpanzee, which must have assisted the animal in moving about in the trees.

By and large, Ardi's foot looks like it was adapted to arboreal climbing and grasping. But team member Owen Lovejoy, an expert on bipedalism, notes that a bone commonly seen in Old World monkeys—the os peroneum, which resides in the tendon of a foot muscle and draws in the great toe during grasping—was found with the Ardi skeleton. Lovejoy postulates that in *Ard. ramidus* this bone would have helped stabilize the foot when the animal walked upright. Interestingly, the os peroneum is not present in modern African apes.

What to make of Ardi's unprecedented amalgam of characteristics? Following their exhaustive analysis of the *ramidus* fossils, particularly Ardi's skeleton, White and his collaborators came to a number of conclusions. First, Ardi was "well-adapted to bipedality," which is to say she could get around on the ground in an upright fashion without teetering side to side like chimps do when they walk on two legs. Ardi may not have been an obligate biped like us, the team acknowledged, but her species did walk upright from time to time, perhaps to carry food. Second, when she was in the trees, Ardi was a careful climber. Third, the forward position of the foramen magnum and shortening of the base of the skull, combined with a reduction in canine size, firmly identifies *Ardipithecus* as a hominid. Fourth, Ardi was not a knuckle-walker. Fifth, she was more monkey-like than chimplike when traveling in the trees. And lastly, Ardi was an ancestor to *Australopithecus* and, therefore, to later hominids, including the lineage that led to modern humans.

If the team is correct, some long-standing notions about human evolution are wrong. Chief among these is the idea that the last common ancestor of humans and chimps was chimplike in many respects. The traditional view is that because hominids and living apes share so much of their anatomy, these features must have been inherited from their last common ancestor, and that ancestor must have been an ape that spent a significant amount of time hanging from branches, as chimps do. Now, however, White and colleagues hypothesize that the last common ancestor looked more like *Ard. ramidus* and traveled on all fours atop the branches like a monkey. This notion flies in the face of more than a century of investigation and implies that many of the similarities between living apes and hominids are not the result of shared ancestry but instead evolved in parallel.

Scholars are reluctant to give up the chimplike model for *ramidus* for a number of reasons. The biggest objection concerns the suggestion that hominids and African apes evolved their similarities independently rather than inheriting them from a common ancestor. Paleontologists construct their phylogenetic trees according to the principle of parsimony: a simpler explanation is always preferred over a more complex one. In this case, the simpler explanation for the similarities between living apes and humans is that they inherited them from a common ancestor. It is less probable (although not impossible) that all those features evolved independently in these groups.

Yet even the discovery team acknowledges that placing *Ard. ramidus* on the human family tree is not a straightforward task. White has offered up several hypotheses for how the species fits in. Two of them propose (in slightly different ways) that *Ard. ramidus* evolved into *Australopithecus* in a single evolving lineage. The third hypothesis suggests that *Ard. ramidus* is not the ancestor to *Australopithecus* and left no later descendants. It is clear, though, that White favors placing the species in one of the ancestral spots.

*   *   *

The paleoanthropology community is only just beginning to digest many of the conclusions about *Ard. ramidus* proffered by White and his collaborators. But already some of their assertions have drawn criticism. Take, for example, the claim that Ardi walked upright when traveling on the ground. William Jungers of Stony Brook University, an expert on the lower limb anatomy of hominids, remarked to ScientificAmerican.com that "based on what [White et al.] present, the evidence for bipedality is limited at best" in *Ard. ramidus*. Ardi's long arms, powerful hands, archless foot, and above all her divergent big toe, all testify against bipedal locomotion.

Only additional fossil finds will help settle the question of whether *Ard. ramidus* walked on two legs. The thigh bone usually contains a wealth of clues about locomotion, but Ardi's is woefully incomplete, missing the critical top and bottom ends. Likewise, a knee joint could reveal whether Ardi walked upright without shifting her weight from side to side in chimpanzee fashion.

If *Ard. ramidus* was not bipedal, would it still qualify as a hominid? Perhaps. But then we would have to redefine what a hominid is. In that case, the defining characteristic would be the presence of a stubby, diamond-shaped upper canine that does not sharpen against the lower premolar the way the canines of other primates do. For me, though, bipedalism is still the ultimate hallmark of humanity, the game-changing innovation that separated our ancestors from the apes.

In fact, looking at Ardi overall, many more traits unite her with apes than with hominids. Jungers has remarked—and I agree—that if we had found only the postcranial bones of Ardi, we would not hesitate to call her an ape. Could this species have been one of nature's experiments that left no descendants? David Begun of the University of Toronto, an authority on Miocene apes, thinks that's a distinct possibility. He told ScientificAmerican.com that he finds "very little in the anatomy of [Ardi] that leads directly to *Australopithecus*, then to *Homo sapiens*," adding that "this could very easily be a side branch." And Chris Stringer of the Natural History Museum in London has made similar comments to the press.

Not every fossil hominid species is an ancestor to later hominids. We know this from the many dead-end branches on the human family tree, including *Homo neanderthalensis, Homo floresiensis,* and several species of *Australopithecus.* White's team argues that *Ard. ramidus* signifies that knuckle-walking evolved independently in chimps and gorillas. But if that's possible, then it's also possible that bipedalism arose more than once. This would mean that just because an ancient creature was bipedal (whether in the trees or on the ground), that doesn't guarantee it a spot on the line that leads to us. Indeed, we already know of a bipedal ape that is not considered ancestral to humans. *Oreopithecus bambolii* is a seven- to eight-million-year-old ape first recognized in 1872 on the basis of fossils from north-central Italy. Recent studies of a more complete skeleton found in 1958 suggest that this European ape, sometimes jokingly called the "cookie ape," may have been largely bipedal. Radiographs of the skeleton's hip blade, or ilium, reveal an internal bone structure that is consistent with bipedality. Furthermore, one of the key features supporting bipedalism in *Ard. ramidus* is a protuberance on Ardi's ilium called the anterior inferior iliac spine (AIIS). Interestingly, the AIIS is also present on the *Oreopithecus* pelvis. Although the *Ard. ramidus* team made only passing reference to *Oreopithecus* in their *Science* papers, more extensive comparisons of Ardi and the cookie ape might be enlightening.

Another reason to question whether *Ard. ramidus* belongs in our lineage is the suggestion that *Ardipithecus* evolved into *Australopithecus. Australopithecus anamensis,* the oldest species of *Australopithecus,* is 4.2 million years old, and based on the parts of the skeleton discovered thus far it looks radically different from the 4.4-million-year-old Ardi. Evolution would have had to work overtime to produce the profound changes needed to make the transition from *Ardipithecus* to *Australopithecus* in just 200,000 years. If it is shown that this is indeed what occurred, scientists will have to explain why human evolution suddenly sped up after millions of years of relative stasis. After all, the putative hominid fossils of the 7-million-year-old *Sahelanthropus* and the 6-million-year-old *Orrorin* (both of which White considers to be older versions of *Ardipithecus*) are considered to be bipedal. Why would hominids have changed so little between 7 million years ago and 4.4 million years ago, and then suddenly morph so dramatically?

One more aspect of the *Ard. ramidus* team's interpretation of their finds warrants discussion here. In the concluding *Science* article, Owen Lovejoy resurrects and elaborates on an argument he made in the same journal in 1981 about the origin of bipedalism and how it led to our global presence, remarkable intellect, unlimited technological skills, and monogamous mating strategy. At the time, he was aiming to explain bipedalism in Lucy's species, *A. afarensis*. Lovejoy sees reduction in canine tooth size as indicative of a significant shift in the life-history strategy of hominids. The idea here is that hominid females chose to mate with males who were less aggressive (ones that lacked the fighting teeth seen in other primates). This led to an increased level of pair bonding that was reinforced by males bringing food back to share with bonded females and offspring. Thus females gave males exclusive mating rights in exchange for food for them and their young. Bipedalism, in this view, evolved as a means of bringing more food back to the females and young because it enabled carrying.

Scientists have objected to Lovejoy's model for the origins of bipedalism, since he first described it in 1981, in large part because unlike *A. afarensis*, modern-day primates that pair-bond do not show dramatic differences in body size between males and females. Furthermore, monogamy would be hard to enforce if males were out gathering food all day away from their home base. Although male and female body sizes are said to be much closer in *Ard. ramidus*, the latest incarnation of Lovejoy's theory does not seem to have won over the critics. Reflecting on the new *Ard. ramidus* research in general, David Pilbeam of Harvard University told the *Washington Post*, "This is an extraordinary achievement, of discovery, recovery, reconstitution, description, and analysis, which will keep many others busy for at least another fifteen years." But when it came to Lovejoy's model, Pilbeam was quoted in the *Los Angeles Times* as saying, "This is a restatement of Owen Lovejoy's ideas going back almost three decades, which I found unpersuasive then and still do."

Kim Hill, a researcher at the Institute of Human Origins who studies hunter-gatherers, recently commented to me that even if *Ardipithecus* males did bring back food for others to share, plant resources are notoriously low in calories. African tubers, a staple for the Hadza hunter-gatherers of Tanzania, have only about 10–50 calories per 100 grams of raw weight. A 1-kilogram root, if it could be digested without cooking, would yield only 100 to 500 calories, depending on what kind of plant it was. Wild fruits and nuts are also relatively low in calories. Providing a family with a sufficient number of calories from plant foods demands transporting very large amounts of vegetables—a feat that would require some kind of a container. Meat is a much richer source of energy: a ten-kilogram mammal, for instance, contains some fifteen thousand calories. But even if *Ard. ramidus* could obtain meat, its reduced canines would not have offered an effective means of ripping open animal skin to access the flesh underneath.

Regardless of where the dust settles on *Ardipithecus ramidus*—and my long experience as a paleoanthropologist tells me that the next decade will see unbridled controversy over this species—this is one astonishing find. Some debates may find resolution once other scholars are permitted to see the original specimens, others may have to await further discoveries from the *Ard. ramidus* time range, and still others may carry on for decades.

The discovery of Ardi, like my discovery of Lucy in 1974, reinforces the value of fieldwork. Such incredible surprises will not be found in museum drawers or university laboratories. Only diligent and persistent fieldwork will bring such rewards.

Selected References

Alemseged, Z., F. Spoor, W. H. Kimbel, R. Bobe, D. Geraads, D. Reed, and J. G. Wynn. "A Juvenile Early Hominid Skeleton from Dikika, Ethiopia." *Nature* 443 (2006): 296–301.

Asfaw, B., T. D. White, C. O. Lovejoy, B. Latimer, S. Simpson, and G. Suwa. "*Australopithecus garhi:* A New Species of Early Hominid from Ethiopia." *Science* 284 (1999): 629–35.

Begun, David R. "The Earliest Hominins—Is Less More?" *Science* 303 (2004): 1478–80.

———. "Planet of the Apes." *Scientific American* (February 2003): 74–83.

———. "Relations Among the Great Apes and Humans: New Interpretations Based on the Fossil Great Ape *Dryopithecus.*" *Yearbook of the American Journal of Physical Anthropology* 37 (1994): 11–63.

Bocherens, Hervé, Dorothée G. Drucker, Daniel Billiou, Marylène Patou-Mathis, and Bernard Vandermeersch. "Isotopic Evidence for Diet and Subsistence Pattern of the Saint-Césaire I Neanderthal: Review and Use of a Multi-source Mixing Model." *Journal of Human Evolution* 49 (2005): 71–87.

Bonnefille, R., R. Potts, F. Chalie, D. Jolly, and O. Peyron. "High-Resolution Vegetation and Climate Change Associated with Pliocene *Australopithecus afarensis.*" *Proceedings of the National Academy of Sciences* 101 (2004): 12125–29.

Brown, P., T. Sutikna, M. J. Morwood, R. P. Soejono, Jatmiko, E. Wayhu Saptomo, and Rokus Awe Due. "A New Small-Bodied Hominin from the Late Pleistocene of Flores, Indonesia." *Nature* 431 (2004): 1055–61.

Brunet, M., A. Beauvilain, Y. Coppens, E. Heintz, A. H. E. Moutaye, and D. Pilbeam. "*Australopithecus bahrelghazali,* une nouvelle espèce d'hominide ancien de la région de Koro Toro (Tchad)." *Comptes rendus des séances de l'Académie des Sciences* 322 (1996): 907–13.

Brunet, Michel, Franck Guy, David Pilbeam, Daniel E. Lieberman, Andossa Likius, Hassane T. Mackaye, Marcia S. Ponce de León, Christoph P. E. Zollikofer, and Patrick Vignaud. "New Material of the Earliest Hominid from the Upper Miocene of Chad." *Nature* 434 (2005): 752–55.

Campisano, C. J., and C. S. Feibel. "Connecting Local Environmental Sequences to Global Climate Patterns: Evidence from the Hominid-Bearing Hadar Formation, Ethiopia." *Journal of Human Evolution* 53 (2007): 515–27.

———. "Depositional Environments and Stratigraphic Summary of the Pliocene Hadar Formation at Hadar, Afar Depression, Ethiopia." In "The Geological Context of Human Evolution in the Horn of Africa," edited by J. Quade and J. Wynn. A Geological Society of America Special Paper, forthcoming.

Clarke, R. J. "First Ever Discovery of a Well-Preserved Skull and Associated Skeleton of *Australopithecus*." *South African Journal of Science* 94 (1998): 460–63.

Darwin, Charles. *The Descent of Man and Selection in Relation to Sex.* 2 vols. London: John Murray, 1871.

———. *On the Origin of Species by Means of Natural Selection, or the Preservation of Favoured Races in the Struggle for Life.* London: John Murray, 1859.

Dennell, Robin, and Wil Roebroeks. "An Asian Perspective on Early Human Dispersal from Africa." *Nature* 438 (2005): 1099–1104.

de Waal, F. B. M. *Bonobo: The Forgotten Ape.* Berkeley: University of California Press, 1998.

———. *Tree of Origin: What Primate Behavior Can Tell Us About Human Social Evolution.* Cambridge, Mass.: Harvard University Press, 2002.

Drapeau, M. "Functional Analysis of the Associated Partial Forelimb Skeleton from Hadar, Ethiopia (A.L. 438-1): Implications for Understanding Patterns of Variation and Evolution in Early Hominin Forearm and Hand Anatomy." Ph.D. diss., University of Missouri–Columbia, 2001.

Drapeau, M., C. V. Ward, W. H. Kimbel, D. C. Johanson, and Y. Rak. "Associated Cranial and Forelimb Remains Attributed to *Australo-*

pithecus afarensis from Hadar, Ethiopia." *Journal of Human Evolution* 48 (2005): 593–642.

Duarte, C., J. Maurmcio, P. B. Pettitt, P. Souto, E. Trinkaus, H. van der Plicht, and J. Zilhão. "The Early Upper Paleolithic Human Skeleton from the Abrigo do Lagar Velho (Portugal) and Modern Human Emergence in Iberia." *Proceedings of the National Academy of Sciences USA* 96 (1999): 7604–9.

Falk, Dean, Charles Hildebolt, Kirk Smith, M. J. Morwood, Thomas Sutikna, Peter Brown, Jatmiko, E. Wayhu Saptomo, Barry Brunsden, and Fred Prior. "The Brain of LB1, *Homo floresiensis.*" *Science* 308 (2005): 242–45; published online March 3, 2005, www .sciencexpress.org.

Gabunia, Leo, Abesalom Vekua, David Lordkipanidze, Carl C. Swisher III, Reid Ferring, Antje Justus, Medea Nioradze, Merab Tvalchrelidze, Susan C. Antón, Gerhard Bosinski, Olaf Jöris, Marie-A.-de Lumley, Givi Majsuradze, and Aleksander Mouskhelishvili. "Earliest Pleistocene Hominid Cranial Remains from Dmanisi, Republic of Georgia: Taxonomy, Geological Setting, and Age." *Science* 288 (2000): 1019–25.

Gibbons, Ann. *The First Human: The Race to Discover Our Earliest Ancestors.* New York: Doubleday, 2006.

————. "Glasnost for Hominids: Seeking Access to Fossils." *Science* 297 (2002): 1464–68.

HaileMichael, M., J. L. Aronson, S. Savin, M. J. S. Tevesz, and J. Carter. "$\delta^{18}O$ Mollusk Shells from Pliocene Lake Hadar and Modern Ethiopian Lakes: Implications for History of the Ethiopian Monsoon." *Palaeogeography, Palaeoclimatology, and Palaeoecology* 186 (2002): 81–99.

Haile-Selassie, Y., G. Suwa, and T. D. White. "Late Miocene Teeth from Middle Awash, Ethiopia, and Early Hominid Dental Evolution." *Science* 303 (2004): 1503–5.

Henshilwood, C. S., F. D'Errico, C. W. Marean, R. G. Milo, and R. J. Yates. "An Early Bone Tool Industry from the Middle Stone Age, Blombos Cave, South Africa: Implications for the Origins of Modern Human Behaviour, Symbolism and Language." *Journal of Human Evolution* 41 (2001): 631–78.

Huxley, Thomas H. *Evidence as to Man's Place in Nature.* London: Williams and Norgate, 1863.

Johanson, Donald C. "Ethiopia Yields First 'Family' of Early Man." *National Geographic* 150 (1976): 791–811.

———. "Face-to-Face with Lucy's Family." *National Geographic* 189 (1996): 96–117.

Johanson, Donald C., and Maitland A. Edey. *Lucy: The Beginnings of Humankind.* New York: Simon and Schuster, 1981.

Johanson, Donald C., and Blake Edgar. *From Lucy to Language.* 2nd edition. New York: Simon and Schuster, 2006.

Johanson, Donald C., and James Shreeve. *Lucy's Child: The Discovery of a Human Ancestor.* New York: William Morrow, 1989.

Johanson, D. C., and T. D. White. "A Systematic Assessment of Early African Hominids." *Science* 203 (1979): 321–29.

Johanson, D. C., T. D. White, and Y. Coppens. "A New Species of the Genus *Australopithecus* (Primates: Hominidae) from the Pliocene of Eastern Africa." *Kirtlandia* 28 (1978): 1–11.

Johanson, D. C. et al. "Pliocene Hominid Fossils from Hadar, Ethiopia." *American Journal of Physical Anthropology* 57 (1982): 373–719.

Kalb, John. *Adventures in the Bone Trade: The Race to Discover Human Ancestors in Ethiopia's Afar Depression.* New York: Copernicus Books, 2001.

Kimbel, W. H., D. C. Johanson, and Y. Rak. "The First Skull and Other New Discoveries of *Australopithecus afarensis* at Hadar, Ethiopia." *Nature* 368 (1994): 449–51.

———. "Systematic Assessment of a Maxilla of *Homo* from Hadar, Ethiopia." *American Journal of Physical Anthropology* 103 (1997): 235–62.

Kimbel, W. H., C. A. Lockwood, C. V. Ward, M. G. Leakey, Y. Rak, and D. C. Johanson. "Was *Australopithecus anamensis* ancestral to *A. afarensis*? A case of Anagenesis in the Hominin Fossil Record." *Journal of Human Evolution* 51 (2006): 134–52.

Kimbel, W. H., Y. Rak, and D. C. Johanson. *The Skull of Australopithecus afarensis.* New York: Oxford University Press, 2004.

Kimbel, W. H., R. C. Walter, D. C. Johanson, K. E. Reed, J. L. Aronson, Z. Assefa, C. W. Marean, G. G. Eck, R. Bobe, E. Hovers,

Y. Rak, C. Vondra, T. Yemane, D. York, Y. Chen, N. M. Evensen, and P. E. Smith. "Late Pliocene *Homo* and Olduwan Tools from the Hadar Formation (Kada Hadar Member) Ethiopia." *Journal of Human Evolution* 31 (1996): 549–61.

Kimbel, W. H., T. D. White, and D. C. Johanson. "Cranial Morphology of *Australopithecus afarensis:* A Comparative Study Based on a Composite Reconstruction of the Adult Skull." *American Journal of Physical Anthropology* 64 (1984): 337–88.

————. "Implications of KNM-WT 17000 for the Evolution of 'Robust' *Australopithecus.*" In *Evolutionary History of the "Robust"* Australopithecines, edited by F. E. Grine, 259–68. New York: Aldine de Gruyter, 1988.

Klein, Richard G., and Blake Edgar. *The Dawn of Human Culture: A Bold New Theory on What Sparked the "Big Bang" of Human Consciousness.* New York: Wiley, 2002.

Larson, Susan G. "Evolutionary Transformation of the Hominin Shoulder." *Evolutionary Anthropology* 16 (2007): 72–87.

Leakey, Meave G., Fred Spoor, Frank H. Brown, Patrick N. Gathogo, Christopher Kiarie, Louise N. Leakey, and Ian McDougall. "New Hominin Genus from Eastern Africa Shows Diverse Middle Pliocene Lineages." *Nature* 415 (2001): 443–44.

Leakey, Meave G., and Alan Walker. "Early Hominid Fossils from Africa." *Scientific American* 276 (1997): 74–79.

Lewin, Roger. "Rock of Ages—Cleft by Laser." *New Scientist* 131 (1991): 36–40.

Lockwood, C. A., W. H. Kimbel, and D. C. Johanson. "Temporal Trends and Metric Variation in the Mandibles and Dentition of *Australopithecus afarensis.*" *Journal of Human Evolution* 39 (2000a): 23–55.

Lordkipanidze, David, Tea Jashashvili, Abesalom Vekua, Marcia S. Ponce de León, Christoph P. E. Zollikofer, G. Philip Rightmire, Herman Pontzer, Reid Ferring, Oriol Oms, Martha Tappen, Maia Bukhsianidze, Jordi Agusti, Ralf Kahlke, Gocha Kiladze, Bienvenido Martinez-Navarro, Alexander Mouskhelishvili, Medea Nioradze, and Lorenzo Rook. "Postcranial Evidence from Early *Homo* from Dmanisi, Georgia." *Nature* 449 (2007): 305–10.

Lordkipanidze, David, Abesalom Vekua, Reid Ferring, G. Philip

Rightmire, Jordi Agusti, Gocha Kiladze, Alexander Mouskhelishvili, Medea Nioradze, Marcia S. Ponce de León, Martha Tappen,
and Christoph P. E. Zollikofer. "The Earliest Toothless Hominin
Skull." *Nature* 434 (2005): 717–18.

Lovejoy, C. O. "Modeling Human Origins: Are We Sexy Because
We're Smart, or Smart Because We're Sexy?" In *The Origin and
Evolution of Humans and Humanness,* edited by D. Tab Rasmussen,
1–28. Boston: Jones and Bartlett, 1993.

———. "The Origin of Man." *Science* 211 (1981): 341–50.

Lovejoy, C. O., R. S. Meindl, J. C. Ohman, K. G. Heiple, and T. D.
White. "The Maka Femur and Its Bearing on the Antiquity of
Human Walking: Applying Contemporary Concepts of Morpho-
genesis to the Human Fossil Record." *American Journal of Physical
Anthropology* 119 (2002): 97–133.

Marean, C. W. "From the Tropics to the Colder Climates: Contrast-
ing Faunal Exploitation Adaptations of Modern Humans and
Neanderthals." In *From Tools to Symbols: From Early Hominids to
Modern Humans,* edited by F. D'Errico and L. R. Backwell, 333–71.
Johannesburg: Witwatersrand University Press, 2005.

———. "Heading North: An Africanist Perspective on the Replacement of Neanderthals by Modern Humans." In *Rethinking the
Human Revolution,* edited by P. Mellars, K. Boyle, O. Bar-Yosef,
and C. Stringer, 367–79. Cambridge: McDonald Institute for
Archaeological Research, 2007.

Marean, C. W., M. Bar-Matthews, J. Bernatchez, E. Fisher, P. Goldberg,
A. I. R. Herries, Z. Zenobia Jacobs, A. Jerardino, P. Karkanas,
T. Minichillo, P. J. Nilssen, E. Thompson, I. Watts, and H. M. William.
"Early Human Use of Marine Resources and Pigment in South Africa
During the Middle Pleistocene." *Nature* 449 (2007): 905–8.

Marzke, M. W. "Joint Function and Grips of the *Australopithecus
afarensis* Hand, with Special Reference to the Region of the Capitate." *Journal of Human Evolution* 12 (1983): 197–211.

Mayr, E. *What Evolution Is.* New York: Perseus Books Group, 2001.

McBrearty, S., and A. Brooks. "The Revolution That Wasn't: A New
Interpretation of the Origin of Modern Human Behavior." *Journal of Human Evolution* 39 (2000): 453–563.

Morwood, Mike, and Penny van Oosterzee. *A New Human: The Startling Discovery and Strange Story of the "Hobbits" of Flores, Indonesia.* New York: Smithsonian Books, 2007.

Morwood, M. J., P. Brown, Jatmiko, T. Sutikna, E. Wahyu Saptomo, K. E. Westaway, Rokus Awe Due, R. G. Roberts, T. Maeda, S. Wasisto, and T. Djubiantono. "Further Evidence for Small-Bodied Hominins from the Late Pleistocene of Flores, Indonesia." *Nature* 437 (2005): 1012–17.

Noonan, James P., Graham Coop, Sridhar Kudaravalli, Doug Smith, Johannes Krause, Joe Alessi, Feng Chen, Darren Platt, Svante Pääbo, Jonathan K. Pritchard, and Edward M. Rubin. "Sequencing and Analysis of Neanderthal Genomic DNA." *Science* 314 (2006): 1113–18.

Pickford, Martin, Brigitte Senut, Dominique Gommery, and Jacques Treil. "Bipedalism in *Orrorin tugenensis* Revealed by Its Femora." *Comptes Rendus Palevol* 1 (2002): 191–203.

Pilbeam, David. Book review in *American Scientist* 66, no. 3 (1978): 378–79.

———. "New Hominoid Skull Material from the Miocene of Pakistan." *Nature* 295 (1982): 232–34.

Potts, Richard, Anna K. Behrensmeyer, Alan Deino, Peter Ditchfield, and Jennifer Clark. "Small Mid-Pleistocene Hominin Associated with East African Acheulean Technology." *Science* 305 (2004): 75–78.

Rak, Y. *The Australopithecine Face.* New York: Academic Press, 1983.

Reed, K. E. "Paleoecological Patterns at the Hadar Hominin Site, Afar Regional State, Ethiopia." *Journal of Human Evolution* 54 (2008): 743–68.

Renne, P., R. Walter, K. Verosub, M. Sweitzer, and J. Aronson. "New Data from Hadar (Ethiopia) Support Orbitally Tuned Time-Scale to 3.3 Ma." *Geophysical Research Letters* 20 (1993): 1067–70.

Richards, Michael P., Paul B. Pettitt, Erik Trinkaus, Fred H. Smith, Maja Paunovic, and Ivor Karavanic. "Neanderthal Diet at Vindija and Neanderthal Predation: The Evidence from Stable Isotopes." *Proceedings of the National Academy of Sciences* 97, no. 13 (2000): 7663–66.

Richmond, Brian G., and William L. Jungers. "Orrorin Tugenensis

Femoral Morphology and the Evolution of Hominin Bipedalism." *Science* 319 (2008): 1662–65.

Roche, Hélène, and Jean-Jacques Tiercelin. "Découverte d'une industrie lithique ancienne in situ dans la formation d'Hadar, Afar central, Éthiopie." *Comptes rendus des séances de l'Académie des Sciences* (Paris) 284 (1977): 1871–74.

Sarich, Vincent M., and Allan C. Wilson. "Immunological Time Scale for Hominid Evolution." *Science* 158 (1967): 1200–36.

Schwartz, Jeffrey H. "Getting to Know *Homo erectus*." *Science* 305 (2004): 53–54.

Schwartz, Jeffrey H., Leo Gabunia, Abesalom Vekua, and David Lordkipanidze. "Taxonomy of the Dmanisi Crania." *Science* 289 (2000): 55–56.

Semaw, S., M. J. Rogers, J. Quade, P. Renne, R. Butler, M. Domínguez-Rodrigo, D. Stout, W. Hart, T. Pickering, and S. Simpson. "2.6-Million-Year-Old Stone Tools and Associated Bones from OGS-6 and OGS-7, Gona, Afar, Ethiopia." *Journal of Human Evolution* 45 (2003): 169–77.

Shreeve, James. *The Neandertal Enigma: Solving the Mystery of Modern Human Origins.* New York: William Morrow, 1995.

Simons, Elwyn L. "Ramapithecus." *Scientific American* 236 (1977): 28–35.

Sloan, C. P. "Origin of Childhood." *National Geographic* 210 (2006): 148–59.

Spoor, F., M. G. Leakey, P. N. Gathogo, F. H. Brown, S. C. Antón, I. McDougall, C. Kiarie, F. K. Manthi, L. N. Leakey. "Implications of New Early *Homo* Fossils from Ileret, East of Lake Turkana, Kenya." *Nature* 448 (2007): 688–91.

Stern, J. T., Jr. "Climbing to the Top: A Personal Memoir of *Australopithecus afarensis*." *Evolutionary Anthropology* 9 (2000): 113–33.

Tattersall, Ian. *Becoming Human: Evolution and Human Uniqueness.* New York: Oxford University Press, 2000.

Tattersall, Ian, Eric Delson, John Van Couvering, and Alison Brooks, eds. *Encyclopedia of Human Evolution and Prehistory.* 2nd edition. New York: Garland, 1999.

Tobias, P. V. *Olduvai Gorge.* Vol. 2, *The Cranium and Maxillary Dentition of* Australopithecus (Zinjanthropus) boisei. Cambridge: Cambridge University Press, 1967.

Tocheri, Matthew W., Caley M. Orr, Susan G. Larson, Thomas Sutikna, Jatmiko, E. Wahyu Saptomo, Rokus Awe Due, Tony Djubiantono, Michael J. Morwood, and William L. Jungers. "The Primitive Wrist of *Homo floresiensis* and Its Implications for Hominin Evolution." *Science* 317 (2007): 1743–45.

Ungar, P. S., and M. F. Teaford, eds. *Human Diet: Its Origin and Evolution.* London: Bergin and Garvey, 2002.

Vekua, Abesalom, David Lordkipanidze, G. Philip Rightmire, Jordi Agusti, Reid Ferring, Givi Maisuradze, Alexander Mouskhelishvili, Medea Nioradze, Marcia Ponce de León, Martha Tappen, Merab Tvalchrelidze, and Christoph Zollikofer. "A New Skull of Early *Homo* from Dmanisi, Georgia." *Science* 297 (2002): 85–89.

Walter, R. C. "Age of Lucy and the First Family: Laser ^{40}Ar/^{39}Ar Dating of the Denen Dora Member of the Hadar Formation." *Geology* 22 (1994): 6–10.

Walter, R. C., and J. L. Aronson. "Age and Source of the Sidi Hakoma Tuff, Hadar Formation, Ethiopia." *Journal of Human Evolution* 25 (1993): 229–40.

White, Tim. "Early Hominids—Diversity or Distortion?" *Science* 299 (2003): 1994–97.

White, T. D., B. Asfaw, D. DeGusta, H. Gilbert, G. D. Richards, G. Suwa, and F. C. Howell. "Pleistocene *Homo sapiens* from Middle Awash, Ethiopia." *Nature* 423 (2003): 742–47.

White, Tim D., Donald C. Johanson, and William H. Kimbel. "*Australopithecus africanus*: Its Phyletic Position Reconsidered." *South African Journal of Science* 77 (1981): 445–70.

White, T. D., G. Suwa, and B. Asfaw. "*Australopithecus ramidus,* a New Species of Early Hominid from Aramis, Ethiopia." *Nature* 371 (1994): 306–12.

Wolpoff, Milford, John Hawks, Brigitte Senut, Martin Pickford, and James Ahern. "An Ape of the Ape: Is the Toumaï Cranium TM 266 a Hominid?" *PaleoAnthropology* (2006): 36–50.

Wong, Kate. "An Ancestor to Call Our Own." *Scientific American* (January 2003): 54–63.

———. "The Littlest Human." *Scientific American* (February 2005): 56–65.

———. "Lucy's Baby." *Scientific American* (December 2006): 78–85.

———. "The Morning of the Modern Mind." *Scientific American* (June 2005): 86–95.

———. "Stranger in a New Land." *Scientific American* (November 2003): 74–83.

———. "Who Were the Neandertals?" *Scientific American* (April 2000): 98–107.

Wrangham, Richard W. "The Significance of African Apes for Reconstructing Human Social Evolution." In *The Evolution of Human Behavior: Primate Models,* edited by Warren G. Kinzey, 51–71. Albany: State University of New York Press, 1987.

Wynn, J. G., R. Bobe, D. Geraads, D. Reed, D. C. Roman. "Geological and Palaeontological Context of a Pliocene Juvenile Hominin at Dikika, Ethiopia." *Nature* 443 (2006): 332–36.

WEBSITES

American Museum of Natural History, www.amnh.org

Atapuerca, www.ucm.es/info/paleo/ata/english/

Darwiniana and Evolution, http://members.aol.com/darwinpage/hominid.htm

Institute of Human Origins, Becoming Human, www.becominghuman.org

Mapping Humanity's Genetic Journey Through the Ages, http://www5.nationalgeographic.com/geographic/index.html

Museum of Paleontology, University of California–Berkeley, www.ucmp.berkeley.edu

National Center for Science Education, www.ncseweb.org

National Geographic Society, www.nationalgeographic.com

Scientific American, www.sciam.com

Smithsonian Institution, www.mnh.si.eduanthro/humanorigins

Talk.Origins, www.talkorigins.org

Acknowledgments

First and foremost I want to thank the Ethiopian people for all of the encouragement, interest, and support they have so graciously offered during my almost forty fruitful and memorable years as a guest in their great nation.

I wish to express my appreciation to the Ethiopian Ministry of Culture and Tourism, particularly to His Excellency, Ambassador Mahmoud Dirir, Minister of Culture and Tourism. Special thanks to Mr. Jara HaileMariam, head of the Authority for Research and Conservation of Cultural Heritage, for granting permission for the Hadar Research Project to conduct research in Ethiopia. The Afar Regional State provided local permissions and offered invaluable support and assistance. At the National Museum of Ethiopia, the staff and the director, Ms. Mamitu Yilma, have provided logistical support for our expeditions and made our laboratory research possible.

Financial support for the Hadar Research Project has come from the National Science Foundation, the National Geographic Society, as well as from supporters and board members of IHO at Arizona State University. I am particularly grateful to the IHO board of directors, which has generously contributed funding for our research. The board of directors, under the able chairmanship of Bruce Schnitzer, continues be a source of support and guidance: Robert Beck, Alejandra Escandon, Thomas F. Hill, Thomas P. Jones III, David H. Koch, Elaine Leventhal, Harry A. Papp, Herb and Laura Roskind, Janet and Ed Sands, Peter Saucier, Ian Tattersall, and Arnold and Joan Travis.

I wish to express my deepest gratitude to my scientific colleagues at IHO: Bill Kimbel, Kaye Reed, Mark Spencer, Gary

Schwartz, Curtis Marean, Lilian Spencer, Kim Hill, and Chris Campisano. I also want to especially recognize the support of IHO staff members Laurel Porter and Bethany Baker, as well as our student workers, José Barrera Flores and Sondra Menzies. The writing of this book has also benefited from discussions with many wonderful IHO/ASU graduate students.

President Michael Crow of ASU and Professor Sander van der Leeuw, director of the School of Human Evolution and Social Change, have encouraged and supported IHO's mission and the writing of this book.

Janet and Ed Sands provided a comfortable environment in which to work on this book at their wonderful home in Santa Barbara, California. For indispensable encouragement and editorial brilliance I wish to thank Robin Stratton. And Gerald Eck, Bill Kimbel, and Zeresenay Alemseged deserve special thanks for their comments on the manuscript.

Since 1990 the Hadar Research Project has benefited from the remarkably committed efforts of:

Geologists: Drs. James Aronson, Kay Behrensmeyer, Craig Feibel, Mulugeta Fesseha, Million HaileMichael, Raymonde Bonnefille, Paul Renne, Carl Vondra, Robert Walter, John Kappelman, and Tesfaye Yemane, Christopher Campisano; Mr. Tony Troutman and Mr. Mike Tesfaye.

Paleontologists/anthropologists: Drs. Gerald Eck, Bill Kimbel, Kaye Reed, Yoel Rak, Rene Bobe, Zeresenay Alemseged, Charles Lockwood, Ray Bernor, Michelle Drapeau, and Elizabeth Harmon; Mr. Mohamed Ahamadin, Mr. Alemayehu Asfaw, Mr. Michael Black, Mr. Tesfaye Hailu, Mr. Ambachew Kebede, Mr. Chalachew Mesfin, Mr. Caley Orr, Ms. Amy Rector, and Mr. Tamrat Wodajo. It was with great sadness that I learned of Charles Lockwood's death during the final stages of writing this book. Charlie was a wonderful colleague and friend and will be missed by everyone who knew him.

Archaeologists: Drs. Erella Hovers and Zelalem Assefa; Ms. Karen

Schollmeyer, Ms. Talia Goldman, Mr. Tadiwos Asebework, Mr. Kebede Geleta, Mr. Tekele Hagos, and Mr. Essayas (Bruk) GebreMariam.

Logistical support in the field was provided by: Alayu Kassa, Mesfin Mekonnen, Getachew Senbeto, Assaye Zerihun, Achamyeleh Teklu, and Abebe Dileligne. The late Wubishet Fantu was a dear friend and member of our camp crew.

My deepest thanks go to the Afar people of Eloaha village: Abdulla Mohamed, Abdu Mohamed, the late Ahmed Bidaru, Ali Mohamed Ware, Ali Welleli, Ali Yussef, the late Dato Adan, the late Dato Ahmedu, Doud Ebrahim, Ebrahim Digra, Ebrahim Habib, Ebrahim Nore, Edris Ahmed, Ese Hamadu, Hamadu Mohamed, Hamadu Humed, Hamadu Meter, Humed Michael, Humed Waleno, Kaloyta Ese, Maumin Alehandu, Meter Dato, Michael Dato, Mohamed Ahmed Bidaru, Mohamed Ese, Mohamed Gofre, Mohamed Omar, Nore Ali, and Omar Abdulla. Mr. Mohamed Ahamadin of the Culture and Tourism Bureau of the Afar Regional State government was central to the HRP's success, as an adviser in all relationships with the Afar people.

We shall miss other late members of the HRP who played critical roles in our success: Meles Kassa, Dato Ahmedu, Dato Adan, and Solomon Teshome.

For his years of valued friendship and generous use of his splendid photographs, I wish to thank Enrico Ferorelli. I wish to express my gratitude to Michael Hagelberg for his excellent illustrations.

The completion of this book would not have been possible without the guidance and encouragement of my literary agent Sterling Lord, whose impact on the world of publishing is legendary. I wish to express my appreciation to our editor John Glusman and others at Harmony Books, who have worked so hard to bring this book to completion.

Collaborations are always fraught with challenge, but in this endeavor my collaboration with Kate Wong has been a memorable pleasure.

No doubt, I have forgotten to acknowledge someone, but if so this is not intentional and I plead your forgiveness. It has been a remarkable experience for me to have worked as a paleoanthropologist, with all of the ups and downs, rewards and disappointments, over the last thirty-eight years. I thank everyone who has encouraged me in my pursuit of the past and look forward to the new generation of paleoanthropologists who will no doubt add immeasurably to understanding our origins.

Kate Wong would like to thank the following people for sharing their paleoanthropological insights through interviews and e-mails: Bill Kimbel, Curtis Marean, Milford Wolpoff, Matt Tocheri, Bill Jungers, Harold Dibble, John Shea, Peter Brown, Mike Moorwood, Marc Meyer, Karen Baab, Svante Pääbo, and David Lordkipanidze. She is also very grateful for encouragement and support from her fellow Scientific Americans.

Index

About the Authors

Pioneering paleoanthropologist and winner of the American Book Award, DONALD C. JOHANSON founded the Institute of Human Origins, now located at Arizona State University in Tempe, in 1981.

KATE WONG has been covering human evolution for *Scientific American* since 1997. She lives in New York City.